HANDBOOK OF
BASIC QUALITY CONTROL TESTS
FOR DIAGNOSTIC RADIOLOGY

The following States are Members of the International Atomic Energy Agency:

AFGHANISTAN	GERMANY	PALAU
ALBANIA	GHANA	PANAMA
ALGERIA	GREECE	PAPUA NEW GUINEA
ANGOLA	GRENADA	PARAGUAY
ANTIGUA AND BARBUDA	GUATEMALA	PERU
ARGENTINA	GUYANA	PHILIPPINES
ARMENIA	HAITI	POLAND
AUSTRALIA	HOLY SEE	PORTUGAL
AUSTRIA	HONDURAS	QATAR
AZERBAIJAN	HUNGARY	REPUBLIC OF MOLDOVA
BAHAMAS	ICELAND	ROMANIA
BAHRAIN	INDIA	RUSSIAN FEDERATION
BANGLADESH	INDONESIA	RWANDA
BARBADOS	IRAN, ISLAMIC REPUBLIC OF	SAINT KITTS AND NEVIS
BELARUS	IRAQ	SAINT LUCIA
BELGIUM	IRELAND	SAINT VINCENT AND
BELIZE	ISRAEL	THE GRENADINES
BENIN	ITALY	SAMOA
BOLIVIA, PLURINATIONAL	JAMAICA	SAN MARINO
STATE OF	JAPAN	SAUDI ARABIA
BOSNIA AND HERZEGOVINA	JORDAN	SENEGAL
BOTSWANA	KAZAKHSTAN	SERBIA
BRAZIL	KENYA	SEYCHELLES
BRUNEI DARUSSALAM	KOREA, REPUBLIC OF	SIERRA LEONE
BULGARIA	KUWAIT	SINGAPORE
BURKINA FASO	KYRGYZSTAN	SLOVAKIA
BURUNDI	LAO PEOPLE'S DEMOCRATIC	SLOVENIA
CAMBODIA	REPUBLIC	SOUTH AFRICA
CAMEROON	LATVIA	SPAIN
CANADA	LEBANON	SRI LANKA
CENTRAL AFRICAN	LESOTHO	SUDAN
REPUBLIC	LIBERIA	SWEDEN
CHAD	LIBYA	SWITZERLAND
CHILE	LIECHTENSTEIN	SYRIAN ARAB REPUBLIC
CHINA	LITHUANIA	TAJIKISTAN
COLOMBIA	LUXEMBOURG	THAILAND
COMOROS	MADAGASCAR	TOGO
CONGO	MALAWI	TONGA
COSTA RICA	MALAYSIA	TRINIDAD AND TOBAGO
CÔTE D'IVOIRE	MALI	TUNISIA
CROATIA	MALTA	TÜRKİYE
CUBA	MARSHALL ISLANDS	TURKMENISTAN
CYPRUS	MAURITANIA	UGANDA
CZECH REPUBLIC	MAURITIUS	UKRAINE
DEMOCRATIC REPUBLIC	MEXICO	UNITED ARAB EMIRATES
OF THE CONGO	MONACO	UNITED KINGDOM OF
DENMARK	MONGOLIA	GREAT BRITAIN AND
DJIBOUTI	MONTENEGRO	NORTHERN IRELAND
DOMINICA	MOROCCO	UNITED REPUBLIC
DOMINICAN REPUBLIC	MOZAMBIQUE	OF TANZANIA
ECUADOR	MYANMAR	UNITED STATES OF AMERICA
EGYPT	NAMIBIA	URUGUAY
EL SALVADOR	NEPAL	UZBEKISTAN
ERITREA	NETHERLANDS	VANUATU
ESTONIA	NEW ZEALAND	VENEZUELA, BOLIVARIAN
ESWATINI	NICARAGUA	REPUBLIC OF
ETHIOPIA	NIGER	VIET NAM
FIJI	NIGERIA	YEMEN
FINLAND	NORTH MACEDONIA	ZAMBIA
FRANCE	NORWAY	ZIMBABWE
GABON	OMAN	
GEORGIA	PAKISTAN	

The Agency's Statute was approved on 23 October 1956 by the Conference on the Statute of the IAEA held at United Nations Headquarters, New York; it entered into force on 29 July 1957. The Headquarters of the Agency are situated in Vienna. Its principal objective is "to accelerate and enlarge the contribution of atomic energy to peace, health and prosperity throughout the world".

IAEA HUMAN HEALTH SERIES No. 47

HANDBOOK OF BASIC QUALITY CONTROL TESTS FOR DIAGNOSTIC RADIOLOGY

ENDORSED BY
THE AMERICAN ASSOCIATION OF PHYSICISTS IN MEDICINE,
THE EUROPEAN FEDERATION OF ORGANISATIONS
FOR MEDICAL PHYSICS AND
THE INTERNATIONAL SOCIETY OF RADIOGRAPHERS
AND RADIOLOGICAL TECHNOLOGISTS

INTERNATIONAL ATOMIC ENERGY AGENCY
VIENNA, 2023

COPYRIGHT NOTICE

© IAEA, 2023

Printed by the IAEA in Austria
February 2023
STI/PUB/2021

IAEA Library Cataloguing in Publication Data

Names: International Atomic Energy Agency.
Title: Handbook of basic quality control tests for diagnostic radiology / International Atomic Energy Agency.
Description: Vienna : International Atomic Energy Agency, 2023. | Series: IAEA human health series, ISSN 2075–3772 ; no. 47 | Includes bibliographical references.
Identifiers: IAEAL 22-01533 | ISBN 978–92–0–130322–6 (paperback : alk. paper) | ISBN 978–92–0–130422–3 (pdf) | ISBN 978–92–0–130522–0 (epub)
Subjects: LCSH: Radiology. | Radiology — Handbooks, manuals, etc. | Radiology — Quality control. | Radiation dosimetry. | Diagnostic imaging.
Classification: UDC 615.849 (035) | STI/PUB/2021

FOREWORD

It is unquestionable that modern medicine would not exist in its present form without X rays, which are widely used for the diagnosis and treatment of patients. In recent years, medical imaging technology has evolved exponentially, shifting steadily from analogue to digital radiology, from single slice to multidetector row computed tomography, and from fluoroscopy to complex and sophisticated angiography systems. During a diagnostic procedure, the physical properties of X rays are used to obtain a diagnosis or to guide invasive devices through the body when an interventional procedure is performed. A simple error or malfunction of an X ray system can affect the health of patients.

This principle necessitates that X ray machines be monitored through strict quality control. Quality control tests provide 'snapshots' of performance that help to ensure optimal performance of X ray equipment in accordance with the clinical objective, thus providing diagnostic information of the required quality with the lowest patient exposure. These tests also assist medical institutions in the responsible management of investment in X ray equipment, as the procurement and maintenance of these systems can account for a large amount of their budget.

In many countries, a significant number of X ray systems used in diagnostic radiology departments are not part of a regular quality assurance programme. This is mainly due to the lack of professionals trained in quality assurance, dosimetry testing and detailed assessment of the performance of X ray systems, as well as the lack of relevant guidance. To address this issue, the IAEA organized a series of technical cooperation projects aimed at building competence in establishing and strengthening quality assurance and quality control in X ray diagnostics and in applying best practices for quality and safety in diagnostic radiology. The present publication is the result of two workshops held in Vienna with the participation of experts representing European countries and representatives of professional societies such as the American Association of Physicists in Medicine, the European Federation of Organisations for Medical Physics and the International Society of Radiographers and Radiological Technologists. The main objective of the workshops was to develop a handbook of harmonized quality control procedures for diagnostic radiology equipment on the basis of existing material, taking into consideration all the latest developments in the field.

This publication has been endorsed by the American Association of Physicists in Medicine, the European Federation of Organisations for Medical Physics and the International Society of Radiographers and Radiological Technologists. The IAEA officers responsible for this publication were H. Delis and V. Tsapaki of the Division of Human Health.

EDITORIAL NOTE

CONTENTS

1. INTRODUCTION

1.1. BACKGROUND

Diagnostic radiology has been established for more than one century and represents one of the most powerful tools used in modern medicine, as effective treatment is closely linked to timely and accurate diagnosis. The performance of ionizing radiation emitting equipment used in medicine needs to be monitored to ensure its safe and effective use [1]. This process starts from basic quality control (QC) and extends to comprehensive quality management systems that integrate every aspect of patient care [2].

QC represents the most basic level of managing quality and includes the set of operations employed to maintain or improve quality [3]. It is considered to provide a snapshot of the performance characteristics of a product or service. QC could be used to verify that the product or the service complies with requirements.

To ensure best practice, more organized efforts are required in the modern era of diagnostic radiology. QC is just one element of a comprehensive quality assurance (QA) programme, which aims to ensure that the quality requirements for a product or service will consistently be fulfilled in every aspect. Operating a comprehensive QA system starts even before the procurement of any equipment, as the assessment of needs and the development of specifications precede purchase. These components require evaluation of how they would fit into the QA framework.

Further expanding on this QA framework, a comprehensive quality management system can provide additional benefits to a diagnostic radiology service. Harmonized policies, procedures and elements such as mission statements and properly prepared job descriptions can provide clarity and consistency in the services provided, enhancing the outcome and providing a solid platform for quality improvement.

The IAEA promotes this comprehensive approach to quality in diagnostic imaging and supports the development of appropriate staff guidance and training in maintaining quality standards. The IAEA technical cooperation projects RER/6/032 (Strengthening Quality Assurance and Quality Control in Diagnostic X rays) and RER/6/028 (Establishing Quality Assurance/Quality Control in X Ray Diagnostics) were aimed at building competence and establishing quality assurance and quality control procedures in X ray diagnostics using best practices for quality and safety in diagnostic radiology. This handbook is based on the work of these technical cooperation projects.

1.2. OBJECTIVE

The aim of this handbook is to provide guidance towards achieving the best possible practices by summarizing detailed information on QC tests found in other publications. This handbook is a compilation of tests that help to reveal key issues in the X ray units used in a diagnostic radiology department and provides a summary of the minimum recommended scope of these tests using a resource stratified approach. For a more elaborate QA programme, it is recommended that other QC procedures — such as more sophisticated QC tests, image reject analysis and artefact identification in patient images — be implemented as well.

Guidance provided here, describing good practices, represents expert opinion but does not constitute recommendations made on the basis of a consensus of Member States.

1.3. SCOPE

This handbook focuses on acceptance tests and routine performance tests, such as status tests and constancy tests. Routine performance tests for radiography, fluoroscopy, angiography, mammography and computed tomography (CT) are described. The tests require only the use of instruments, reproducible and known objects (phantoms) and test objects, which do not cost much. The handbook provides a quick reference guide on how to perform each test and raises attention to common issues and mistakes that could undermine the results or the evaluation of a given test. Readers are encouraged to revise the QC procedures listed and introduce more extensive tests in their own department or to adjust the frequency of the tests according to previous experience and the stability of a given X ray unit.

1.4. STRUCTURE

This handbook is divided into sections, each dealing with a given imaging modality. Section 1 is an introduction to the subject. Section 2 outlines QC tests on radiography, Section 3 on fluoroscopy and angiography, Section 4 on mammography and Section 5 on CT. Section 6 describes tests that are common for all modalities, and Section 7 presents tests that are specific to film–screen systems. Finally, the Appendix provides summary tables of the tests.

The tests are categorized according to the qualified personnel that is responsible to perform them — either qualified medical radiation technologists (radiographers) or medical physicists. Since QA is a team effort, it is strongly recommended that a team of experienced staff including a radiologist, a radiographer and a medical physicist introduce and implement the QA programme in their own department.

2. RADIOGRAPHY

This section describes the requirements for routine QC tests for computed radiography and digital radiography X ray systems. The latter is also referred to as digital radiography or direct digital radiography. The main objective of these tests is to verify the operational stability of the equipment. It is assumed that acceptance testing and commissioning tests are performed and baseline values are established. The procedures described for the following tests are indicative and serve as guidelines to the qualified personnel performing them. Several of the following tests can also be performed on fluoroscopy equipment.

2.1. QUALITY CONTROL TESTS FOR RADIOGRAPHERS

2.1.1. X ray–light beam alignment and centring

2.1.1.1. Description and objective

Improper beam alignment and centring will impact the radiographic image. The set light field needs to align well with the X ray beam area in order to limit the radiation field to the necessary size and to not miss any parts because of possible misalignments. Therefore, the objective of this test is to ensure the coincidence and alignment of the collimated light field with the X ray field. Another aspect is the coincidence of the crosshairs of the collimated light beam with the centre of the X ray beam, which is the origin point of the image [4, 5].

2.1.1.2. Equipment

The following equipment is used for the tests:

(a) An appropriate test object for the alignment of the X ray beam and the light field. Coins or other metal objects may also be used for the alignment test.

(b)   A cylinder with an attenuator at its centre, or another test object that can be used to assess the perpendicular incidence of the beam on the image receptor.

2.1.1.3.   Procedure

The following procedures are used to perform the X ray–light field alignment and centring tests using a commercial test object or coins:

(a)   Using a commercial test object:
   (i)   Place the image receptor on a flat surface and set the X ray tube's axis to be perpendicular to the image.
   (ii)   Position the test object on the image detector with the cylinder in the centre of the light beam crosshairs (Fig. 1).
   (iii)   Set the source to image distance (SID) to be 100 cm. A different SID may be used as well, but it is easier to interpret the results if 100 cm is used.
   (iv)   Use the light field and the markings on the test object for accurate field alignment.
   (v)   Make an exposure using exposure parameters 50 kV and 3 mA · s or the values recommended by the medical physicist.

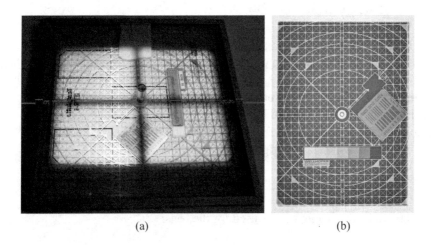

(a)   (b)

FIG. 1. (a) Set-up for X ray–light beam alignment and centring; (b) X ray image of the test object used for X ray–light beam alignment and centring.

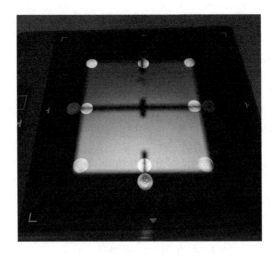

FIG. 2. Set-up for X ray–light beam alignment using coins.

(b) Using coins:
   (i) Place the image receptor on a flat surface and set the X ray tube's axis to be perpendicular to the image.
   (ii) Position the image detector at an SID of 100 cm.
   (iii) Collimate the light beam to a small field on the image detector (e.g. 20 cm × 20 cm). The image receptor area needs to be greater than the light field.
   (iv) Place metal markers (e.g. coins) at every edge of the light field; mark with coins the corner of the light field (Fig. 2).
   (v) Make an exposure using exposure parameters 50 kV and 3 mA · s or the values recommended by the medical physicist.

2.1.1.4. Analysis and interpretation

The following procedures are used for the analysis and interpretation of the test results:

(a) X ray–light field alignment:
   Measure the distance between the edges of the image and the markers of the test object or the coins.
(b) Centring:
   Evaluate the deviation of the light beam crosshairs from the centre of the X ray beam according to the specifications of the test object's manufacturer.

2.1.1.5. Baselines and tolerances

The following baselines and tolerances are used:

(a) X ray–light field alignment:
Ensure that the deviation of the X ray field from the light field does not exceed ±2 cm on any side for an SID of 100 cm or ±2% of any other SID.
(b) Centring:
Ensure that the coincidence of the collimator light beam crosshairs with the X ray beam centre does not exceed ±1 cm at SID = 100 cm.

2.1.1.6. Frequency

Repeat the test every three to six months.

2.1.1.7. Corrective actions

If the tolerance specified in Section 2.1.1.4 is exceeded, the following actions should be taken:

(a) Check the geometry of the measurement settings and repeat the test.
(b) If a misalignment persists, ask the medical physicist to perform a more thorough investigation. Fluorescent screens (from an old cassette), coins or other markers and the use of a mobile phone with video recording functionality may provide an inexpensive solution for testing. Alternatively, self-developing films could be used.
(c) If a trend is observed when repeating this test, then the service engineer should adjust the beam limiting device.
(d) If misalignment occurs in a random manner, then check for observable slack.

2.1.2. Distances and scales

2.1.2.1. Description and objective

The aim of this test is to determine the accuracy of the SID indicator of the X ray system. This test has an important role for radiographic imaging and when dose monitoring software is used, as the distance indication may be used as an input for determining the distance from the focal spot to the entrance surface (skin) [4].

2.1.2.2. Equipment

The equipment needed for the test is a long ruler or a measuring tape at least 100 cm long.

2.1.2.3. Procedure

The following procedure is used for the test:

(a) Set an SID of 100 cm by using the SID indicator of the system.
(b) Measure the actual SID using the ruler or measuring tape, starting at the focal spot indicator on the side of the X ray tube and extending to the tabletop or surface of the image receptor (Fig. 3).
   Note: Some recent X ray systems do not have an indicator of the focal spot. In this case, the manufacturer's documentation could be used as a reference to estimate its position, or the medical physicist may determine the location of the focal spot using the intercept theorem.
(c) Record the result.

2.1.2.4. Analysis and interpretation

Compare the measured value with the SID readout indicator of the X ray system.

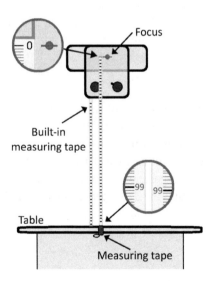

FIG. 3. Schematic of the suggested set-up.

2.1.2.5. Baselines and tolerances

The difference between the ruler or tape measurement and the SID readout indicator should not exceed ±1.5 cm.

2.1.2.6. Frequency

The test is repeated every six months.

2.1.2.7. Corrective actions

If the tolerance is exceeded, then the following actions must be taken:

(a) Repeat the test.
(b) Correct systematic errors. If a digital display of the SID is in use, then the service engineer needs to calibrate the display. If only a tape measure is available, then the correction should be calculated on a case by case basis.
(c) Check for slack in the measuring tape. The service engineer should fix this.

2.1.3. Image uniformity and artefacts

2.1.3.1. Description and objective

This test is performed to assess the degree and source of artefacts visualized in digital radiography images, and to ensure that the image is uniform and artefact free [4].

2.1.3.2. Equipment

A copper attenuator (1 mm thick or similar) is recommended. Alternatively, other attenuators may be used, such as 10 cm of polymethyl methacrylate (PMMA). However, use needs to be consistent and at least the same materials with the same properties should be used if the attenuator itself is not the same. The attenuator should be large enough to cover the whole image, as well as homogeneous. Figure 4 shows the described arrangement.

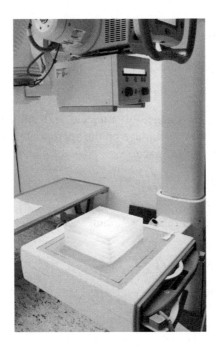

FIG. 4. Set-up for image uniformity test using PMMA.

2.1.3.3. Procedure

The test procedure is as follows:

(a) Place the attenuator in the X ray field.
(b) Set an SID of 100 cm.
      Note: For computed radiography systems, a larger SID (at least 150 cm) and two exposures may be necessary in order to assess whether the lack of uniformity is attributable to the heel effect. After making the first exposure, turn the image receptor around its axis, perpendicular to the image plane, by 180° and make a second exposure. Then, compare the two images.
(c) Place the copper plate at the collimator or place the attenuator at a distance where it covers the whole image receptor area.
(d) Set the collimator in a way that the entire image receptor is exposed.
(e) Set the exposure parameters at 70 kV and 3 mA · s or at the recommended values set by the medical physicist; then make an exposure.

2.1.3.4. *Analysis and interpretation*

The following procedure is used for the analysis and interpretation of the test results:

(a) Visually inspect the image for artefacts and uniformity. Use a narrow window width to view the image on the appropriate display.
(b) Record the status of the image.

2.1.3.5. *Baselines and tolerances*

No visible artefacts or grossly inhomogeneous areas should be observable.

2.1.3.6. *Frequency*

The test is repeated monthly.

2.1.3.7. *Corrective actions*

For computed radiography plates and films, spots indicate that the plate needs to be cleaned, while a line indicates that the reader's optical elements need maintenance or cleaning. Cracks at the edge of the image may suggest that the plate should be removed from service and replaced.

It is not necessary to check all the cassettes at the same time, but there should be a routine to ensure that all cassettes are checked on a regular basis.

For digital radiography systems, lines and rectangular areas in the image indicate that recalibration of the detector is required or that clusters of dead pixels are present. With clusters of dead pixels possibly undermining clinical diagnosis, the detector may need to be replaced; this should be discussed with the service engineer. The radiologists evaluating images acquired with this X ray machine should be made aware of the observed issue and consulted on use of the system.

2.1.4. Automatic exposure control constancy

2.1.4.1. *Description and objective*

The purpose of automatic exposure control (AEC) is to deliver consistent, reproducible exposures across a wide range of tube potentials and to compensate for different anatomical thicknesses. The consistency of the AEC device is assessed with this test [4, 6].

2.1.4.2. Equipment

A copper plate or other attenuator, as described in Section 2.1.3, is used for this test.

2.1.4.3. Procedure

The test procedure is as follows:

(a) Centre the tube on the image receptor area.
(b) Collimate the beam so that it covers the whole image receptor. The field size needs to be set in a way it covers the AEC sensor(s).
(c) Put the attenuator on the X ray tube or place it at the appropriate distance between the focal spot and the image receptor so that it covers the whole image receptor area (Fig. 5).
(d) Select the exposure parameters according to the local routine or use the parameters applied during the commissioning procedure (e.g. 70 kV). Choose the AEC sensor designated or used earlier for this test. The tube kilovoltage should be kept the same for all measurements.
(e) Make an exposure and record the tube current–exposure time product ('mAs value') and the exposure indicator (exposure index) displayed on the operator's console.

FIG. 5. Arrangement to check the constancy of the automatic exposure control sensor using a copper attenuator.

2.1.4.4. Analysis and interpretation

Compare the mAs value and exposure index with the corresponding baseline values.

2.1.4.5. Baselines and tolerances

The recorded exposure index and mAs values should be within ±25% of the respective baseline values for digital radiography systems and within ±30% for computed radiography systems.

2.1.4.6. Frequency

The test is repeated every three months.

2.1.4.7. Corrective actions

If the tolerance is exceeded, then proceed as follows:

(a) Repeat the test, in order to exclude the possibility of measurement error.
(b) If the problem persists, ask the medical physicist to perform a more thorough investigation.

Since the operation of the AEC depends on the generator and AEC system itself, the consistency of the generator should be checked. Trends indicate that a recalibration of the AEC system by the service engineer may be necessary.

2.1.5. Condition of cassettes and image plates (computed radiography only)

2.1.5.1. Description and objective

To ensure that computed radiography plates are clean and have no mechanical defects, which may lead to image artefacts, the following conditioning process may be used. There are no references available in the literature for this test; however, the supplier's user manual may be used. No cleaning solutions or methods should be used other than those explicitly mentioned in the supplier's user manual for the computed radiography cassettes.

2.1.5.2. Equipment

Cleaning solution and cloth, according to the supplier's maintenance manual, are required for this test.

2.1.5.3. Procedure

The following procedure is followed:

(a) Perform a visual examination for external defects of each cassette. Pay special attention to hinges and locks (Fig. 6).
(b) Each cassette should be opened and the imaging plate removed to inspect for dust and scratches (Fig. 7). Record the results of the inspection.
(c) Cleaning of the imaging plates should be performed according to the supplier's maintenance manual.
(d) Erase all the cassettes to ensure that plates are not left for a significant period of time without being used or erased.

FIG. 6. Defective cassette in which the lock on the right cannot be closed.

FIG. 7. Defective cassette in which scratches on the screen prevent the processing unit to process it properly.

2.1.5.4. Analysis and interpretation

It is recommended that this procedure be performed on all the imaging plates on a monthly basis and whenever a number of small white speck artefacts is observable in the images obtained from an imaging plate. Note that too frequent cleaning of the imaging plates or the use of a non-approved cleaning solution can discolour the phosphor.

2.1.5.5. Baselines and tolerances

There should be no dirt or damage to the imaging plates.

2.1.5.6. Frequency

The tests are repeated according to the supplier's recommendations or monthly if artefacts are seen in the images.

2.1.5.7. Corrective actions

When artefacts or defects are observed which may impact clinical use of the cassettes then they should be thoroughly cleaned according to the supplier's instructions and checked to see if the cassette can still be used. Otherwise the cassette should be replaced.

2.1.6. Automatic exposure control sensitivity

2.1.6.1. Description and objective

As described in Section 2.1.4.1, AEC is used to compensate for the change of the X ray tube voltage and the thickness of the patient's anatomical area exposed. All radiography systems equipped with AEC use AEC sensors, which are usually ionization chambers that determine the amount of photons incident on the image receptor. Consistent image quality requires that each sensor produce the same output (or at least a consistently different setting if it is set differently) for a given setting and operate in a consistent manner [4, 6].

2.1.6.2. Equipment

For this test a copper plate or other attenuator is used, as described in Section 2.1.3.

2.1.6.3. Procedure

The test procedure is as follows:

(a) Centre the tube to the image receptor placed in the bucky.

(b) Collimate the X ray beam to cover the whole area of the attenuator and ensure that the selected AEC ionization chamber is inside the X ray beam.

(c) Slide the copper plate into the X ray tube accessory rail or place the attenuator between the focal spot and the bucky.

(d) Select 70 kV and the central ionization chamber (Fig. 8). The kilovoltage should be kept the same whenever this test is carried out, along with other AEC settings, such as film speed or density correction, if they are selectable. Their values should be set to default or normal level (speed: 400 (D); density correction: 0.0).

(e) Make an exposure and record the mAs and exposure index values displayed on the X ray system after the exposure. Then select the sensor or sensors that were not tested, and repeat the previous step.

2.1.6.4. Analysis and interpretation

Compare the mAs and exposure index values with the corresponding baseline values.

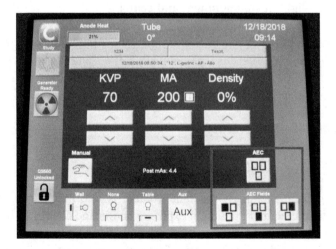

FIG. 8. Control system used for the selection of automatic exposure control ionization chambers (here, the central ionization chamber is selected).

The exposure index and mAs values should not differ by more than ±50% from the corresponding baseline values.

2.1.6.6. Frequency

The test is repeated every month to every three months, depending on the observed consistency of the system.

2.1.6.7. Corrective actions

If the limiting values are exceeded, repeat the test in order to exclude the possibility of error with the set-up. If the limiting values are consistently exceeded, check whether the AEC sensors are set in a different way than expected. The medical physicist should investigate whether the X ray generator performance is adequate. Whenever a defect of the AEC system is observed, avoid its use until it is repaired. A common fault is that the image receptor is not set appropriately for the given X ray unit. Follow the recommendations outlined in Sections 2.2.17–2.2.19 and consider suspending or limiting the use of the system.

2.2. QUALITY CONTROL TESTS FOR MEDICAL PHYSICISTS

2.2.1. X ray–light beam alignment and centring

The test is as described in Section 2.1.1 [4, 5]. The supervising medical physicist should perform this test annually. The same tolerances apply. If a significant difference from previous measurements is found, then the methodology and the performance of this test should be reviewed.

2.2.2. Tube potential accuracy

2.2.2.1. Description and objective

In diagnostic radiology, the tube voltage is one of the most important parameters that affect both radiation exposure and image contrast. The accuracy of tube voltage is critical, since even a minor variation will have considerable effect on the final radiographic image. This test assures that the accuracy of the tube voltage selected from the operator's console of the X ray system corresponds to the actual value [4].

2.2.2.2. Equipment

A solid state kilovolt meter with a valid calibration and capability of measuring in the 40–150 kV range is used. Whenever the tube potential accuracy is measured, the same indicator of the instrument (e.g. practical peak voltage) or the average or maximum of the peak tube voltage) should be consistently used. The choice of indicator may influence the results slightly. A lead plate or apron of thickness 0.35 mm, 0.5 mm or 1.0 mm is needed if the image receptor is not removable from the bucky.

2.2.2.3. Procedure

The following procedure is used:

(a) Remove the image receptor or cover it with a lead plate if it is not removable.
(b) Place the instrument on a flat surface with its sensitive area facing the X ray tube. Ensure that the X ray beam is perpendicular to and centred on the instrument.
(c) Set the distance between the focal spot and the instrument at 100 cm. This is not mandatory; however, it is convenient to use this setting for the calculation of the radiation output later.
   Note: Some instruments require that a check or position check is done prior to the first exposure. Refer to the user manual of the instrument's manufacturer.
(d) It is recommended to position the kilovolt meter so that its active area is oriented perpendicular to the anode–cathode axis of the X ray tube and to keep the field size as small as possible (e.g. 5 cm × 8 cm) to minimize scattering. Use the light field's crosshairs to position the detector in the centre of the X ray field (Fig. 4).
   Note: Ensure that no additional filtration is selected where not applicable.
(e) Select a tube current–time product that does not overload the X ray tube but that provides sufficient dose for a reliable measurement (e.g. 40 mA · s with half of the maximum permitted tube current).
(f) Measure the manually set tube voltage and record the results at least at five clinically relevant settings (e.g. 60, 70, 80, 100 and 120 kV).
   Note: Some instruments can measure multiple parameters (e.g. kilovoltage, dose, time, half-value layer (HVL)) of the same exposure.

2.2.2.4. Analysis and interpretation

Calculate the deviation of measured kilovoltage values from the nominal ones.

2.2.2.5. Baselines and tolerances

The deviation of the measured kilovoltage values from the nominal ones should be within ±5% or ±5 kV, whichever is greater. These values are set as remedial levels in the referenced literature, meaning that action should be taken when such deviations are observed. However, Ref. [4] and the International Electrotechnical Commission (IEC) recommend a suspension level of ±10% or ±10 kV, whichever is greater. In the case of ±10% or ±10 kV, the system should not be used until corrective actions have been taken.

2.2.2.6. Frequency

The test is repeated annually.

2.2.2.7. Corrective actions

If the tolerances are exceeded, check the set-up, positioning and calibration of the kilovolt meter and repeat the test. If the instrument can measure the generator's kilovoltage and dose waveform, then observe them; line voltage supply ripples may affect the measurements or an electrical fault may be present. If the problem persists after repeating the test, then contact the supplier of the X ray system.

2.2.3. Radiation output consistency

2.2.3.1. Description and objective

The radiation output (mGy/(mA · s)) of the radiographic X ray system at a specific tube voltage should remain constant when a given current–time product is selected in any combination of current and time. This test monitors the effect of changes in tube current, time and their product (mAs value) on the radiation output. Furthermore, it is required that the radiation output of the radiographic system have a linear relationship with the current or time selected (e.g. without changing the tube current) but when doubling the exposure time, the radiation output should be double as well.

Some X ray systems, especially mobile radiographic X ray units, can set only the current–time product. In such a case, only the mAs linearity of the system may be tested [4, 7].

2.2.3.2. Equipment

A solid state detector or an appropriate ionization chamber calibrated for radiographic beam qualities is used. A lead plate is necessary if the image receptor is not removable from the bucky.

2.2.3.3. Procedure

(a) Remove the image receptor or cover it with a lead plate if it is not removable.
(b) Place the instrument on a flat surface with its sensitive area facing the X ray tube. Ensure that the X ray beam is perpendicular to and centred on the instrument.
(c) If an ionization chamber is used, then it should be placed so that no backscatter affects the measurement. If the back of the instrument is shielded, then placing it closer to the focal spot is not necessary.
(d) Set the distance between the focal spot and the instrument at 100 cm. This is not mandatory; however, it is convenient to use this setting for the calculation of the radiation output.
    Note: Some solid state detectors require that a check or position check is done prior to the first exposure. Refer to the user manual of the instrument's manufacturer.
(e) It is recommended to position the solid state detector so that its active area is oriented perpendicular to the anode–cathode axis of the X ray tube and to keep the field size as small as possible (e.g. 5 cm × 10 cm) to minimize scattering and to ensure a narrow beam geometry. Use the light field's crosshairs to position the detector in the centre of the X ray field (Fig. 9).
    Note: Ensure that no additional filtration is selected where not applicable and that the appropriate calibration factors are used where applicable.
(f) Use the manual exposure mode and select 80 kV. Choose exposure parameters that result in a constant mAs value (see Table 1 for example settings).

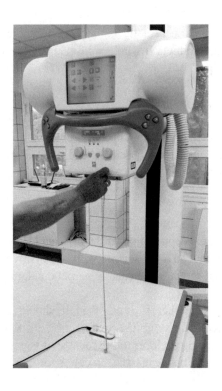

FIG. 9. Set-up for kilovoltage accuracy test using a solid state non-invasive kilovolt meter.

TABLE 1. EXAMPLES OF CONSTANT CURRENT–TIME COMBINATIONS

Current (mA)	Time (s)	mAs value (mA · s)
320	0.125	40
160	0.25	40
80	0.5	40

Note: Alternatively, the linearity of the system may be checked if a time is set to be constant (e.g. 100 ms) and the tube current is changed for each exposure (e.g. 100 mA, 200 mA, 400 mA and 800 mA). Whenever linearity is checked, current could also be kept constant (e.g. 100 mA) for a series of exposures where the exposure time is doubled each time (e.g. 10 ms to 3200 ms).

2.2.3.4. Analysis and interpretation

Calculate the radiation output (μGy/(mA · s)) for all exposures, using Eq. (1):

$$Y = \frac{M_{\mathrm{c}}}{Q} \times \left(\frac{d_{\mathrm{FDD}}}{d_{\mathrm{ref}}}\right)^2 \tag{1}$$

where

Y is the radiation output (mGy/(mA·s));
M_{c} is the corrected reading of the instrument (μGy);
Q is the charge or current–time product (mA·s);
d_{FDD} is the focal spot to instrument distance (m);

and d_{ref} is the reference distance (m; in this case, $d_{\mathrm{ref}} = 1$ m).

2.2.3.5. Baselines and tolerances

The following baselines and tolerances are used:

(a)   Deviations should be within ±20% of the baseline determined during commissioning as a remedial level and ±50% as a suspension level;
(b)   The radiation output for 2.5 mm Al total filtration and 80 kV exposures should be in the range of 25 μGy/(mA · s) to 80 μGy/(mA · s).
      Note: Total filtration is an important factor when one considers the radiation output. The criteria should be adapted when a larger filtration thickness is used. The upper value of the given range could be even higher for some X ray systems that perform satisfactorily.

2.2.3.6. Frequency

The test is repeated annually.

2.2.3.7. Corrective actions

If a consistent deviation is observed from the reference value, then recalibration of the generator may be necessary. If the deviation is not consistent, then an electrical fault may be present in the generator. To check whether the generator is working properly, one should perform the other tests described in this section (i.e. short term reproducibility of the radiation output, kilovoltage accuracy and HVL). Furthermore, the test of total filtration may be performed to investigate whether any additional filter settings remain selected in the X ray beam from previous exposures.

2.2.4. Short term reproducibility of radiation output and exposure time

2.2.4.1. Description and objective

This test evaluates the short term reproducibility of the generator. Although in a clinical environment most systems equipped with AEC are almost exclusively used in this mode, this test may reveal defects of the generator when issues are observed in the AEC system [4, 5, 7].

2.2.4.2. Equipment

The test requires a dosimeter with a solid state detector or an appropriate ionization chamber calibrated for radiographic beam qualities, capable of measuring exposure time in the range of 5 ms to 5000 ms. If the instrument used is not capable of measuring the exposure time, then the dose rate and the integral dose of the exposure may be used to estimate it. This method is less reliable.

It is important to emphasize the importance of the calculation method for the exposure time, as solid state detectors may be set to distinguish different levels of the kilovoltage or dose curves, which may significantly impact their readings.

A lead plate is necessary if the image receptor is not removable from the bucky.

2.2.4.3. Procedure

The following procedure is used:

(a) Remove the image receptor or cover it with a lead plate or lead apron if it is not removable.
(b) Place the instrument on a flat surface with its sensitive area facing the X ray tube. Ensure that the X ray beam is perpendicular to and centred on the instrument.
(c) If an ionization chamber is used, then it should be placed at a distance of at least 20 cm from the table so that no backscatter affects the measurement. If the detector is back shielded, then placing it closer to the focal spot is not necessary.
(d) Set the distance between the focal spot and the instrument at 100 cm. This is not mandatory; however, it is convenient to use this setting for the calculation of the radiation output.

Note: Some solid state detectors require that a check or position check is done prior to the first exposure. Refer to the user manual of the instrument's manufacturer.

(e)   It is recommended to position the solid state detector so that its active area is oriented perpendicular to the anode–cathode axis of the X ray tube and to keep the field size as small as possible (e.g. 5 cm × 10 cm) to minimize scattering and to ensure a narrow beam geometry. Use the light field's crosshairs to position the detector into the centre of the X ray field (Fig. 9).
       Note: Ensure that no additional filtration is selected where not applicable, and that the appropriate calibration factors are used where applicable.

(f)   Use the manual exposure mode and select 80 kV and a current–time combination that results in an exposure of about 40 mA · s (e.g. 400 mA and 100 ms).

(g)   Make five exposures using the same settings and record the results.

2.2.4.4. Analysis and interpretation

The analysis and interpretation of the test results is carried out as follows:

(a)   Calculate the radiation output using Eq. (1) for all measurements, and determine their mean value;

(b)   Calculate the coefficient of variation (COV) for all measured exposure times.

2.2.4.5. Baselines and tolerances

The baselines and tolerances are as follows:

(a)   The deviation of the measured radiation output should be within ±20% of the mean value;

(b)   The COV of the exposure time should be less than ±5%.

2.2.4.6. Frequency

The test is repeated annually.

2.2.4.7. Corrective actions

If after repeated tests the deviation persists, check the tube voltage and radiation dose waveform, if these are available. Potential electrical or electronic failures should be checked by a service engineer.

2.2.5. Exposure time accuracy

2.2.5.1. Description and objective

The generator should be capable of terminating the exposure after a preselected time interval. This test monitors the consistency of the exposure time according to the nominal values set at the console of the X ray system [4].

2.2.5.2. Equipment

An instrument capable of measuring exposure time in the range 5–5000 ms is required. Short exposure times and the calculated exposure times may be affected by the sensitivity settings of the instrument. Refer to the manufacturer's user manual. A lead plate is necessary if the image receptor is not removable from the bucky.

2.2.5.3. Procedure

The following procedure is used:

(a) Remove the image receptor or cover it with a lead plate if it is not removable.
(b) Place the instrument on a flat surface with its sensitive area facing the X ray tube. Ensure that the X ray beam is perpendicular to and centred on the instrument.
(c) Set the focal spot's distance from the instrument at 100 cm (Fig. 9).
    Note: Some solid state detectors require that a check or position check is done prior to the first exposure. Refer to the user manual of the instrument's manufacturer.
(d) Use the manual exposure mode and select 80 kV and 200 mA.
(e) Make at least five different exposures, selecting the times most frequently used in the clinical settings, and record the results.

2.2.5.4. Analysis and interpretation

Calculate the deviation between the nominal and measured values of the exposure time.

2.2.5.5. *Baselines and tolerances*

The following baselines and tolerances are used:

(a) For exposure times longer than 100 ms, ±10% of the nominal value;
(b) For exposure times shorter than 100 ms, ±15% or ±2 ms of the nominal value, whichever is greater.

2.2.5.6. *Frequency*

The test is repeated annually.

2.2.5.7. *Corrective actions*

If the tolerance is exceeded, repeat the test. If a consistent deviation is observed, then check the sensitivity settings of the instrument and observe whether the waveforms are consistent for the same exposure settings, if the instrument is capable of recording waveforms.

2.2.6. Half-value layer

2.2.6.1. *Description and objective*

The aim of this test is to measure the HVL and to confirm that there is sufficient filtration in the X ray beam to remove low energy radiation, according to the minimum requirements of national and international standards [6, 8, 9].

2.2.6.2. *Equipment*

The following equipment is required:

(a) Ionization chamber or solid state detector calibrated for radiographic beam qualities;
(b) High purity (≥99.9%) aluminium attenuators;
(c) Measuring tape;
(d) Metal plate to shield the image receptor from X rays (a lead plate or apron can be used, as mentioned previously), large enough to cover the active area of the detector if the image receptor cannot be removed from the bucky;
(e) Holder for the ionization chamber.

2.2.6.3. Procedure

The test procedures are as follows:

(a) When using aluminium attenuators:
  (i) Place shielding on the image receptor if it cannot be removed from the bucky or move the X ray beam into a position where it cannot irradiate the image receptor (free exposure mode), to protect the image receptor from excessive exposure that could cause artefacts.
  (ii) Set the voltage at 80 kV and select a current–time product (e.g. 40 mA · s) that provides a sufficient dose at the dosimeter.
  Note: Make sure that no additional filtration is selected where not applicable.
  (iii) Place the detector at an appropriate height (1 m or 50 cm) above the image receptor so as to avoid backscatter, centred in the radiation field. If the back of the instrument is shielded, then placing it closer to the focal spot is not necessary.
  (iv) Collimate the radiation field to cover the sensitive area of the dosimeter.
  (v) Make an exposure without an attenuator and record the readings (Fig. 10).
  (vi) Place a sufficiently thick aluminium attenuator (either at the exit of the X ray beam or between the focal spot and the instrument, if a holder is available). The aluminium attenuator should be placed so that it covers the active area of the detector, and its thickness corresponds to the expected HVL (e.g. to measure an expected HVL of 2.5 mm, use an aluminium attenuator of thickness $\geq$2.0 mm, close to the expected value).
  (vii) Make an exposure with the same parameters as before.
  (viii) Record the results and repeat the previous steps, increasing or decreasing the thickness of aluminium attenuators until the two dose values fall below and above 50% of the unattenuated value.
  Note: It is recommended that at the time of commissioning, the HVL measurements are carried out with appropriately calibrated ionization chambers of higher accuracy and precision. Other detectors should only be used for constancy measurements.
(b) When using a solid state detector capable of measuring the HVL:
  (i) Place shielding on the image receptor if it cannot be removed from the bucky or move the X ray beam into a position where it cannot irradiate the image receptor (free exposure mode), to protect the image receptor from excessive exposure that could cause artefacts.

Note: The image receptor should be removed or covered by a lead plate if it is not removable.

(ii) Set the voltage at 80 kV and use a fixed mAs value (e.g. 40 mAs) that gives a sufficient dose at the instrument.

Note: Make sure that no additional filtration is selected where not applicable.

(iii) Place the meter on the patient holder and align it with the beam; then, set the instrument to measure and display the HVL.

(iv) Make three exposures and record the results (HVL).

FIG. 10. Set-up with an ionization chamber for measuring air kerma without attenuators.

2.2.6.4. Analysis and interpretation

The analysis and interpretation of the test results are carried out as follows:

(a) When using aluminium attenuators:
    (i) Calculate the mean value for the three exposures without the attenuator in the beam.
    (ii) Plot a semilogarithmic graph of the measured results (measured dose versus the thickness of the attenuator in millimetres) and find the thickness of the aluminium attenuator required to reduce the unattenuated beam by 50%. The value found by this method is the HVL. Alternatively, the HVL can be calculated with the two-point method using Eq. (2):

$$\mathrm{HVL} = \frac{t_2\ln(2M_1/M_0) - t_1\ln(2M_2/M_0)}{\ln(M_1/M_2)} \tag{2}$$

where

    M_0 is the corrected value of the instrument reading taken without any attenuators;

    M_1, M_2 is the corrected value of the instrument reading closest to half of the unattenuated yield;

and t_1, t_2 is the thickness of the aluminium attenuator (in millimetres) used to determine the corresponding results.

(b) When using a solid state detector capable of measuring the HVL:
Calculate the average of the three exposures.

2.2.6.5. Baselines and tolerances

The HVL needs to comply with the minimum values specified in the national regulations. If no such requirements are available in the regulations, then international and national standards could be applied [6, 8, 9]. The standard IEC 60601-1-3 recommends that at 80 kV, the first HVL for X ray systems marketed before 1 June 2012 be not less than 2.3 mm Al; for systems marketed after that date, the recommended value is 2.9 mm Al [8]. Further recommended first HVL values for different tube voltages may be found in Ref. [6].

2.2.6.6. Frequency

The test is repeated annually.

2.2.6.7. Corrective actions

If the measured HVL does not meet the specified national minimum requirements, then repeat the test and check the instrument and the geometry of the set-up and ensure that the attenuators are appropriate. If these tests confirm the discrepancy, then check the accuracy of the measured kilovoltage (Section 2.2.2) and radiation output (Sections 2.2.3 and 2.2.4). If the HVL is too low, then placement of additional non-removable filtration may be necessary, although this may result in accelerated ageing of the X ray tube. If the HVL is higher than expected, check for removable additional filtration in the X ray beam.

2.2.7. Kerma–area product meter accuracy

2.2.7.1. Description and objective

The kerma–area product (KAP) is used to describe patient dose. The KAP can be determined by multiplying the air kerma on the central axis of the X ray beam with the beam area perpendicular to this axis, at the same distance from the focal spot. This test is undertaken to ensure the accuracy of the KAP meter, which gives an indication of the entrance patient dose. Some X ray systems do not have a KAP meter but have an indicator to inform the operator about the radiation exposure. In all cases, the accuracy of the KAP measurement can be assessed with this test (i.e. verification of readings) [9–11].

2.2.7.2. Equipment

The following equipment is used for this test:

(a) Dosimeter with calibrated ionization chamber or solid state detector;
(b) Computed radiography plate or self-developing film (Fig. 11) or other image receptor calibrated to perform measurements of the field size;
(c) Ruler or measuring tape;
(d) Lead plate if the image receptor is not removable.

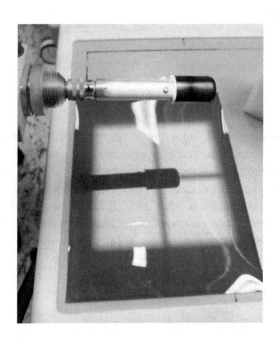

FIG. 11. Set-up using an ionization chamber and a self-developing film to determine the accuracy of the kerma–area product meter.

2.2.7.3. Procedure

The test procedure is as follows:

(a) Remove the image receptor or cover it with a lead plate if it is not removable.
(b) Place the reference instrument on the central axis of the X ray beam at least 20 cm above the couch (tabletop) to avoid the influence of backscattered radiation (for a back shielded solid state detector, this is not needed). A block of extruded polystyrene foam on a holder/stand can be used to support the instrument.
(c) Collimate the X ray beam so that it covers an area of approximately 10 cm × 10 cm on the sensitive area of the reference instrument. Do not change the field size throughout this measurement.
(d) Make manual exposures, irradiating the reference instrument and the KAP meter simultaneously using combinations of tube voltage and total filtration that are typically used in clinical applications.
Note: In the case of fluoroscopy, the most frequently used imaging protocol should be used and noted. When automatic exposure rate control (AERC) is needed to determine the exposure parameters, it may be necessary to cover

the image receptor with a copper plate or to place a phantom below the reference instrument to obtain a higher exposure rate. Backscatter should be considered if the ionization chamber used is sensitive to it. For fluoroscopy, depending on the arrangement of the reference instrument, the attenuation of the table may affect the measurement.

(e)   Record the measurements from the KAP meter and the ones obtained by the reference instrument.

(f)   Remove the reference instrument and position an appropriate image receptor perpendicular to the central axis of the X ray beam at the same distance from the focal spot.

(g)   Expose the image receptor to determine the field size, using low exposure parameters (e.g. 50 kV and 4 mA · s).

2.2.7.4.   Analysis and interpretation

The analysis and interpretation of the test results are carried out as follows:

(a)   Measure the exact field size using a ruler or using an appropriate image display and computer programme to determine the beam area at the position of the detector during the exposure. If a computer programme is used, then the settings of the electronic distance measuring tools need to be calibrated appropriately.

(b)   Calculate the measured KAP by multiplying the results recorded by the reference instrument with the beam area and estimate the deviation between the measured and displayed values. Make sure that all units obtained through the measurements are corrected and converted appropriately when comparing the readings of the KAP meter and the corrected readings from the reference instrument.

2.2.7.5.   Baselines and tolerances

It is expected that the KAP meter's uncertainty be not higher than ±25% [10]. If the system has an indicator instead of an individual KAP meter, then its accuracy should be better than ±35% [11].

2.2.7.6.   Frequency

The test is repeated annually.

2.2.7.7. Corrective actions

If the KAP meter's uncertainty is higher than ±25%, verify the corrections, the calibration factors and the consistency of the units of measure used for the comparison of the readings of the reference instruments and the KAP meter. If the deviation from the reference instrument is constant, then a correction factor can be introduced. If the deviation is outside the tolerances or is inconsistent, then a new calibration may be necessary. If the performance of the other elements of the X ray unit is adequate, the unit may be used until the KAP meter is repaired.

2.2.8. Short term reproducibility of exposure indicator

2.2.8.1. Description and objective

Digital radiography and computed radiography systems have a wide response, which enables them to accommodate a wide exposure range and provide a visually acceptable image even in the case of under- or overexposure. The exposure indicator gives an indication of the absorbed dose at the detector and can be considered as the digital equivalent of optical density for film–screen systems. The exposure index does not always have a linear relationship with the detector dose, since different manufacturers define it differently. When evaluating test results of the exposure index, always refer to the user manual of the given system. Because the exposure index is used to test the AEC, it needs to be linearized when there is no linear relationship with the air kerma measured by the detector. The linearized exposure index can be found using the inverse relationship between the air kerma measured by the detector and the exposure index. At commissioning, the accuracy of the exposure index should be checked using the manufacturer's protocol.

In digital radiography, it is essential to perform routine monitoring of the exposure index. The objective of this test is to confirm that there have been no substantial long term changes in the exposure index of the image receptor. Note that this test requires a baseline to be set — if possible, at the time of commissioning [4, 12].

2.2.8.2. Equipment

The equipment used is a copper plate or other attenuator, as described in Section 2.1.3.

2.2.8.3. Procedure

The test procedure is as follows:

(a) Arrange the set-up so that the X ray tube is directed to the image receptor and centred. An SID of 100 cm is appropriate.
(b) Manually select the exposure parameters (e.g. 70 kV).
(c) Adjust the collimation so that the X ray beam coincides with the image receptor.
(d) Slide the copper plate to the exit window of the collimator or place it in the beam so that it covers its whole area.
(e) Choose a mAs value that delivers 10 µGy to the image receptor.
    Note: For computed radiography plates, it is important to read the plate after a set time period following the exposure (e.g. 1 min). For this measurement, apply a linear readout algorithm.
(f) Make at least three exposures and record the exposure indices of the measurements.

2.2.8.4. Analysis and interpretation

Compare the exposure indices recorded with the initial baseline values for the same exposure parameters and estimate their deviation from the baselines.

2.2.8.5. Baselines and tolerances

The deviation of the exposure index from the baseline value needs to be less than ±10%. If deviations are observed, then the errors should be confirmed by dose measurements.

2.2.8.6. Frequency

The test is repeated annually.

2.2.8.7. Corrective actions

If the tolerance is exceeded, then repeat the test to ensure that the same geometry and exposure parameters are used as during the commissioning. If the failure persists, then measure the detector air kerma and compare it with that measured during commissioning. For a computed radiography system, the reader may have to be cleaned or recalibrated to compensate for the drift of the system.

If the X ray machine performs adequately, then it could be used while ensuring that it will be checked as soon as possible.

2.2.9. Exposure indicator accuracy

2.2.9.1. Description and objective

This test monitors the accuracy of the exposure index for a series of exposures. For this test, the determination of a baseline value is essential [4, 6, 12].

2.2.9.2. Equipment

The equipment required is a copper plate or other attenuator, as described in Section 2.1.3.

2.2.9.3. Procedure

The following procedure is used:

(a) Arrange a set-up similar to that described in Section 2.2.8.3, in which the X ray tube is directed at the image receptor and centred. An SID of 100 cm is appropriate.
(b) Adjust the collimation to have the X ray beam coincide with the image receptor.
(c) Slide the copper plate at the exit window of the collimator or place it in the beam so that it covers its whole area.
(d) Manually select a tube voltage (e.g. 70 kV).
(e) For a computed radiography plate, read the plate after a set time period (e.g. 1 min) and apply a linear readout algorithm.
(f) Expose the image receptor once using a mAs value that delivers about 10 µGy. Then make two more exposures with two different mAs values (a fraction — e.g. 20% — and a multiple — e.g. 10 times — of the initial mAs value) to obtain a wider range of exposure indices.

2.2.9.4. Analysis and interpretation

Calculate the COV for the five measurements.

The deviation of the exposure index from the baseline value needs to be less than ±20%.
Note: If deviations are observed, then the errors should be confirmed by dose measurements.

2.2.9.6. Frequency

The test is repeated annually.

2.2.9.7. Corrective actions

If the tolerance is exceeded, then repeat the test to ensure that the same geometry and exposure parameters are set as during the commissioning. If the failure persists, then measure the detector air kerma and compare it with the one measured during commissioning. For a computed radiography system, the reader may have to be cleaned or recalibrated to compensate for the drift of the system. If the change in exposure index over consecutive tests follows a trend (e.g. a continual increase or decrease in value), discuss the results with the service engineer of the supplier.

If the X ray machine performs well otherwise, then it could be used, but the issue should be investigated as soon as possible.

2.2.10. Image receptor dose

2.2.10.1. Description and objective

This test is used to estimate the image receptor dose that is determined by the AEC system. The test result is system specific and needs to be compared with the baseline values determined during the commissioning and calibration of the AEC. Numerous parameters can be used during the calibration of AEC system (e.g. detector air kerma, signal difference to noise ratio (SDNR), pixel value, exposure index) and the measurement geometries can be with or without scatter. In this test, the image receptor air kerma is used to estimate the AEC dose in a scatter free geometry. However, it is important to set the geometry and parameters used during commissioning [4, 6].

2.2.10.2. Equipment

The following equipment is used in the test:

(a)   Copper plate or other attenuator, as described in Section 2.1.3 (Fig. 12);
(b)   Dosimeter with calibrated ionization chamber or solid state detector.

2.2.10.3. Procedure

The following procedure is used:

(a)   Centre the X ray beam on the central AEC ionization chamber and use the collimator to confine the beam to this sensor.
(b)   Remove the grid and the KAP meter if they are present and removable; otherwise leave them in place and record this fact.

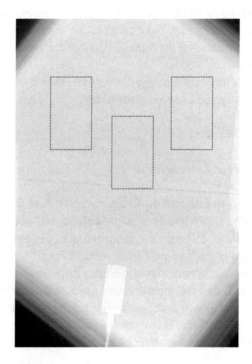

FIG. 12. X ray image showing the set-up for measuring the image receptor dose. A solid state detector and the AEC ionization chamber positions during a measurement under PMMA plates are indicated.

(c)   Place the instrument on the cover of the image receptor centrally within the X ray beam. If it is not possible to put the instrument directly on the image receptor surface, then place it between the source and the image receptor to avoid backscatter (this is not necessary for solid state multimeters).

(d)   Place the copper plate in front of the X ray beam.

(e)   Select the minimum exposure parameters used in clinical practice.

(f)   Make an exposure and record the reading; then apply the appropriate corrections.

(g)   Repeat the previous steps and select the average and the maximum exposure parameters used in clinical practice.

2.2.10.4. Analysis and interpretation

The analysis and interpretation of the test results are carried out as follows:

(a)   Calculate the image receptor dose using Eq. (3):

$$D_{IR} = D_M \times T_{PR} \times \frac{r^2}{SID^2} \qquad (3)$$

where

D_{IR} is the dose to the image receptor;

D_M is the corrected instrument reading (considering calibration factor, temperature, pressure and every further conversion factor necessary);

T_{PR} is the transmission factor, which is necessary only if the grid has not been removed, and it can be found in the documentation of the manufacturer (otherwise $T_{PR} = 1$);

r is the distance between the X ray source and the reference point of the instrument;

and SID is the distance between the focal spot and the image receptor.

(b)   Compare the calculated image receptor dose values with the baseline values.

2.2.10.5. Baselines and tolerances

The deviation of the image receptor dose from the baseline value should be within ±30%.

2.2.10.6. Frequency

The test is repeated annually.

2.2.10.7. Corrective actions

If the tolerance is exceeded, then repeat the test to ensure that all the arrangements correspond to those set during commissioning. Perform a thorough investigation into the possible sources of error. A systematic deviation may indicate a faulty AEC that requires recalibration. A change in the image receptor dose may indicate the fault of the X ray generator or the X ray tube. Test the radiation output using the tests described in Sections 2.2.2–2.2.4 to confirm that the X ray generator and the tube are not faulty. If the AEC needs recalibration, then its use is limited to manual exposure mode.

2.2.11. Leakage radiation

2.2.11.1. Description and objective

This test is designed to determine the leakage radiation of the collimator system and the X ray tube assembly. It is particularly important to carry out this test on mobile X ray units and on any system after any service intervention involving the housing or the collimator system [13].

2.2.11.2. Equipment

The following equipment is used in the test:

(a) A survey meter with a pressurized ionization chamber, or an ionization chamber calibrated for air kerma rate for X ray beam qualities used in diagnostic radiology. An instrument calibrated for ambient dose equivalent rate may be suitable with the appropriate corrections introduced, if necessary.
(b) A 4 mm thick lead sheet covering the exit window of the collimator.
(c) A ruler or a measuring tape.

2.2.11.3. Procedure

The test procedure is as follows:

(a) Arrange the X ray tube so that it is accessible from any direction at 1 m distance. If the tube cannot be arranged in such a way, then note the positions where it is usually used and the directions that it usually points to during clinical practice (Fig. 13).

(b) The X ray parameters of the system should correspond to the highest possible settings. Systems where the AERC determines the exposure parameters should be set to the highest values permitted by the imaging protocol.

Note: It is not necessary to use the maximum permitted exposure parameters (to spare the tube and the personnel from unnecessary load). The leakage radiation could also be calculated by selecting the highest possible kilovoltage and a fraction of the maximum mAs value at that setting (e.g. 150 mA · s for a maximum of 600 mA · s at 150 kV), as this exposure would be sufficient for the purpose of the measurement.

(c) Collimate the beam to the smallest possible size.

Note: Several X ray units do not permit an exposure while the collimator is completely shut. In this case, the 4 mm thick lead sheet by itself is sufficient to shield the direct radiation.

(d) Place the lead sheet on the exit window of the collimator or its assembly to cover it completely. If the equipment is fluoroscopic, a lead apron should be worn when performing measurements around the X ray tube assembly

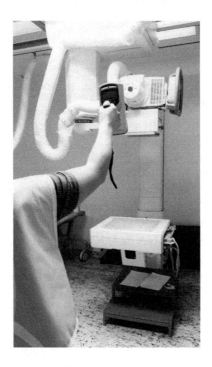

FIG. 13. Measurement of leakage radiation.

at 1 m distance from the focal spot. If the equipment is radiographic only, it is only possible to measure in integration mode and estimate the exposure over 1 h using the set and the maximum allowed mAs values.

Note: The number of measurements carried out under the scope of this test should include at least the anode and cathode sides of the tube and, as in the most frequently used arrangement, any direction in which personnel might be located during exposure.

(e) Record the measured values, the exact sampling direction and the distance from the focal spot.

2.2.11.4. Analysis and interpretation

The dosimeter readings are corrected for the measured quantity if necessary and then compared with the limits. The maximum permitted current rating should be used for the estimation.

2.2.11.5. Baselines and tolerances

The maximum air kerma should not be more than 1 mGy in 1 h at 1 m in any direction.

2.2.11.6. Frequency

The test is performed at the time of acceptance and after major changes to the collimator or the tube assembly.

2.2.11.7. Corrective actions

Repeat the measurement and ensure that the settings are appropriate. Check for any service interventions affecting the X ray tube assembly, including the beam limiting device. The manufacturer may have different specifications for the maximum settings; therefore, refer to the documentation to find the appropriate X ray parameters. A more rigorous examination could be carried out using films or self-developing films by covering the tube's side where the highest measured results were encountered. It is not advisable to use the equipment with a faulty beam limiting device.

2.2.12. Scattered radiation

2.2.12.1. Description and objective

This test is performed to evaluate the occupational exposure of staff present in the room during clinical procedures. The measurement of scattered radiation is more closely related to radiation protection of the staff and to the evaluation of the shielding of the room. This test should be performed on systems that may be used in fluoroscopic mode [14].

2.2.12.2. Equipment

The following equipment is needed:

(a)   A survey meter;
(b)   A 25 cm thick water phantom, or PMMA slabs with a total thickness of 25 cm, large enough to cover the image receptor;
(c)   A measuring tape.

2.2.12.3. Procedure

The test procedure is as follows:

(a)   Set the arm holding the X ray tube and the image receptor to be vertical. The shielding around the table may be arranged as it is used in clinical practice.
(b)   Set the parameters of the system to the most frequently used settings in clinical practice; alternatively, use the settings in the manufacturer's documentation for the distribution map of stray radiation.
       Note: Fluoroscopy X ray units operate in AERC mode by default. In this case, the selected imaging protocol determines the additional filtration(s) and the exposure parameters; therefore, these should be documented as well.
(c)   Choose the largest field size and open the collimator to ensure the largest possible field size, covering the whole area of the phantom.
(d)   Place the phantom on the table and move it as close as possible to the image receptor. Make notes of the geometrical arrangement and the settings.
(e)   While wearing a lead apron, perform dose rate measurements around the X ray tube assembly at the points of interest, including the designated areas for the operators and staff (Fig. 14).
(f)   Record the measured values and their exact sampling direction and distances.

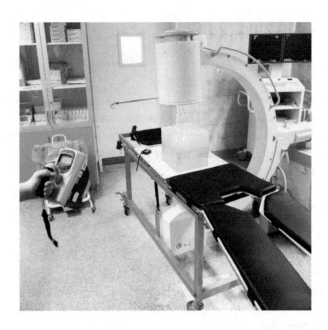

FIG. 14. Set-up for the measurement of scattered radiation around a mobile C arm fluoroscopy unit in the area occupied by staff.

2.2.12.4. Analysis and interpretation

The measured dose rates should be corrected if necessary and then compared with the baseline values.

2.2.12.5. Baselines and tolerances

A difference of ±50% from the baseline value is acceptable. The manufacturer supplies a map of stray radiation in the system documentation, to which compliance could be checked, but this usually does not account for the effect of any shielding which may be present.

2.2.12.6. Frequency

This test is repeated annually.

2.2.12.7. Corrective actions

Repeat the test after checking the exposure parameters, the calculations and the set-up of the geometry, especially if the arrangement of the shielding has changed. If the measured values are not acceptable, then further shielding or a new shielding arrangement may be necessary. Depending on the nature of the change in the scattered radiation, it may be necessary to limit or suspend work until adequate radiation protection tools are provided.

2.2.13. Low contrast detectability

2.2.13.1. Description and objective

The aim of this test is to check the low contrast detectability by using a simple test object on a regular basis. Performance is evaluated simply by counting the number of low contrast details visible on the test object; however, a more elaborate image analysis is suggested. An ongoing record of these numbers will reveal any trend towards deterioration in imaging performance. This test requires a baseline to be set (if possible, at commissioning) [12, 15].

2.2.13.2. Equipment

The following equipment is required:

(a) Low contrast test object (examples in Fig. 15).
   Note: Refer to the test object manufacturer's user manual for its specifications and possible further recommendations.
(b) Copper plate or other attenuator as described in Section 2.1.3.

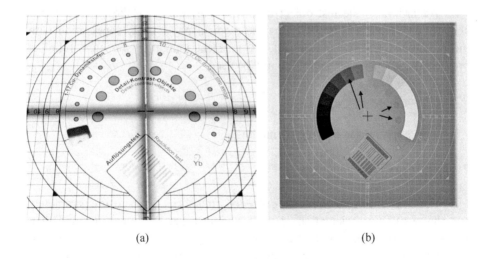

(a)   (b)

FIG. 15. Images of a low contrast test object: (a) phantom; (b) X ray image.

2.2.13.3. Procedure

The test procedure is as follows:

(a)   Position the low contrast test object on the table as close to the image receptor as possible.
(b)   Set a distance of the focal spot to the image receptor of about 100 cm.
(c)   Collimate the X ray beam to cover the whole test object.
(d)   Put the copper plate on the tube assembly.
(e)   If computed radiography is used, erase the plate before the exposure.
(f)   Make an exposure according to the test object's manufacturer specifications (e.g. 70 kV with AEC) or use the parameters determined during commissioning.

2.2.13.4. Analysis and interpretation

Evaluate the image on the display used for diagnostic purposes. Adjust the windowing and magnification to optimize the visibility of the details. Count the number of resolvable low contrast objects on the image, record it and compare it with the baseline. The test is simple and based on a subjective visual check, and it should reveal trends towards the deterioration of low contrast imaging performance.

2.2.13.5. Baselines and tolerances

According to the test object used, the tolerance from the baseline value should be decided and proposed by the medical physicist in cooperation with the radiologist and the radiographer. Consistent imaging parameters (clinical protocols) and image processing should be used.

2.2.13.6. Frequency

The test is repeated every six months.

2.2.13.7. Corrective actions

If the determined tolerance is exceeded, first check the exposure parameters and the positioning/orientation of the test object and the attenuator, and then repeat the test. Changing low contrast visibility may be due to a change of imaging protocols, a change in the arrangement of the measurement set-up or a change in the radiation output and the HVL. Check these to ensure that imaging is consistent. Ageing image receptors, especially computed radiography plates, may also cause deterioration of the low contrast visibility. Consult with staff evaluating images from this system to ensure that the change in visibility does not affect clinical imaging. Further actions may involve the limitation of use or the suspension from service, depending on the level of image quality produced by the X ray machine.

2.2.14. Limiting spatial resolution

2.2.14.1. Description and objective

One of the performance characteristics of the image receptor is its spatial resolution. To characterize this property of the system, one may use a simple test object such as a line pattern resolution test object. The spatial resolution is measured by simply evaluating which groups of line pairs are visible on the image. An ongoing record of these numbers will reveal any trend towards deterioration in imaging performance. This test requires a baseline to be set (preferably during commissioning) [4].

2.2.14.2. Equipment

A spatial resolution (bar pattern) test object is used for this test (an example is given in Fig. 16).

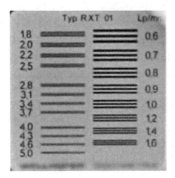

FIG. 16. Image of bar pattern used for the visual determination of the limiting spatial resolution.

2.2.14.3. Procedure

The test procedure is as follows:

(a) Position the test object on the image receptor. If this is not possible, then lay the test object on the table or the bucky and rotate it at approximately 45° to the image receptor edges. In this way interference can be avoided, and the possible highest resolution is observable with this method.
(b) If computed radiography is used, be sure to erase the plate before performing the test.
(c) Make an exposure according to the test object's manufacturer specifications. If no such recommendation is available, then use a sufficiently high exposure to avoid noise on the image. For test objects a few millimetres thick, 50 kV and 10 mA · s may be appropriate; however, for test objects incorporating a few centimetres of PMMA, it may be necessary to increase the exposure parameters.
(d) Record whether or not a grid was used during the test.
(e) Perform this test for two focal spot sizes and every clinically relevant image processing mode (if available).
Note: If this test is performed with a phantom (e.g. 20 cm PMMA), a higher tube voltage (70–80 kV) is necessary.

2.2.14.4. Analysis and interpretation

View the image on a display used for diagnosis. Adjust the windowing and magnification to optimize visualization of the test object. Check which groups of line pairs are visible on the phantom. The line pairs need to be visible over the full length of each group to pass as 'visible'. The last group in which each bar is visible as a distinct line indicates the limiting spatial resolution (in line pairs per millimetre) of the system. Look up the corresponding spatial resolution in test object's specifications and compare with the previous value.

2.2.14.5. Baselines and tolerances

There should be no change in spatial resolution from the baseline. Ensure that the observing conditions are similar to those used earlier.

2.2.14.6. Frequency

The test is repeated every six months.

2.2.14.7. Corrective actions

If there is suspected deterioration of the image resolution, then repeat the test. Check the beam quality and the selected focal spot size, along with the grid. A lower than expected resolution may be due to the viewing conditions or to the subjective nature of the test. Refer to the latest test to check the image quality and take corrective actions to rule out the possible sources of the observed change.

2.2.15. Dark noise

2.2.15.1. Description and objective

This test characterizes another aspect of the image receptor. It is used to assess the level of noise inherent in the readout system; that is, the noise arising from the system itself, without a signal. This test requires a baseline to be set (if possible, this should be done during commissioning) [4, 12, 16].

2.2.15.2. Equipment

Lead sheets or a lead apron (only in the case of digital radiography systems) are needed for the test.

2.2.15.3. Procedure

The test procedure is as follows:

(a) Centre the X ray tube on the image receptor and then cover it with a lead apron or the lead sheets (Fig. 17).
(b) For computed radiography systems, use a freshly erased cassette and leave it in position for about 5 min.
(c) If a digital radiography system is used, obtain an image with a very low exposure to trigger the image receptor readout (e.g. 40 kV and 1 mA · s).
    Note: The collimator may be used as well as a radio opaque absorber. However, some X ray systems cannot make an exposure if a collimator is used and an error message may appear on the console.
(d) For computed radiography systems, use the manufacturer's guidelines to process the plate.

FIG. 17. Digital radiography image receptor covered with a lead apron to test dark noise.

2.2.15.4. Analysis and interpretation

Process the image and draw a region of interest (ROI) in the central part of the image (e.g. 10% of the size of the imaging plate). Measure the mean pixel value (MPV) and record the exposure index.

In addition, perform a visual evaluation and observe the image. A clear, uniform image free of artefacts needs to be obtained when viewed with the clinically used window width and level settings.

2.2.15.5. Baselines and tolerances

Ensure that the recorded exposure index and MPVs do not differ by more than ±50% from the baseline values.

2.2.15.6. Frequency

The test is repeated annually.

2.2.15.7. Corrective actions

For computed radiography systems, repeat the erasure cycle and the test. If the problem persists, check for light leaks in the cassette or the reader and then request service support. Excessive dark noise may also be indicative of laser power loss.

If a digital radiography system is under test, a rise in the MPV or the exposure index may indicate that a recalibration is required for offset correction.

If the dark noise of the system cannot be compensated, the image quality should be discussed with users of the system regarding whether this error affects clinical image evaluation. If recalibration or the introduction of corrections do not resolve this issue, then replacement of the image receptor should be considered. Discuss the results with the service engineer.

2.2.16. Accuracy of measured dimensions

2.2.16.1. Description and objective

The image formed on the image receptor is often used to measure or scale sizes and distances of some findings. This test may be used to ensure that the distances measured from the image by the software distance indicators correspond to the actual distances [4, 12].

2.2.16.2. Equipment

An attenuating test object of known dimensions or a lead ruler is needed for the test.

2.2.16.3. Procedure

The test procedure is as follows:

(a) Centre the X ray beam on the image receptor.
(b) Set the distance of the focal spot to the image receptor at about 100 cm and measure it.
(c) Place the test object on the image receptor or as close to it as possible, in order to minimize magnification. The arrangement should be such that the distance measurements can be performed in the centre and at the periphery of the image receptor. Depending on the object, more exposures may be needed.
(d) Make an exposure using a low tube potential and a sufficient mAs value to visualize the test object with an acceptable contrast. Depending on the object, more exposures may be needed.

2.2.16.4. Analysis and interpretation

Display the image on the monitor used for diagnosis. Measure the dimensions of the test object using the measurement tools on the viewing station, and compare the measurement results with real distances (Fig. 18). Use corrections if magnification has to be used.

2.2.16.5. Baselines and tolerances

The measured dimensions need to be within ±2%, and preferably within ±1%, of the expected values.

2.2.16.6. Frequency

The test is repeated annually.

2.2.16.7. Corrective actions

Repeat the test and make sure that the rulers in the software are positioned accurately on the image of the test object. An incorrect setting in the software may also cause errors in the measurements (e.g. pixel size not corresponding to the real value). Check the user adjustable settings. If the problem persists, request service support. Users of the system should be made aware if the dimensions are

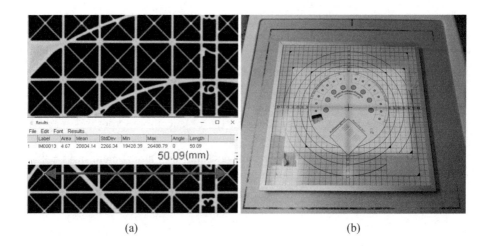

(a)   (b)

FIG. 18. (a) Set-up; and (b) evaluated X ray image of the test object.

not displayed correctly, and service support should be requested to calibrate the measured dimensions.

2.2.17. Automatic exposure control system — consistency between sensors

2.2.17.1. Description and objective

The AEC device may use more than one sensor to determine the necessary exposure for the object being imaged. The consistency between sensors and various combinations of sensors on a single AEC system can be checked with this test. It is very important to emphasize that not every system's sensors have the same sensitivity, as, by design, some systems (e.g. for chest screening) are specialized to a given imaging task [4, 12].

2.2.17.2. Equipment

The equipment needed for the test is a suitable attenuator (e.g. PMMA slabs, a water filled tank, other homogeneous phantoms that cover the whole AEC sensor area). The attenuator should have sufficient thickness to trigger an appropriate response from the AEC system (e.g. 15 cm).

2.2.17.3. Procedure

The following procedure is used:

(a) Centre the tube on the image receptor, which is situated in the bucky. Set the SID at the grid focusing distance and ensure that the grid is in place.
(b) Place the attenuator on the table and centre it so that it covers the AEC sensors.
(c) Collimate the beam to include all AEC sensors.
(d) Select the central AEC sensor and a tube voltage of 70 kV.
(e) Make an exposure and record the post-exposure mAs value and exposure index.
(f) Repeat the same procedure for all the possible combinations of the AEC sensors: central; left; right; left and right; left and central; central and right; and left and central and right AEC sensors (Fig. 19). Use the same kilovoltage for all exposures.

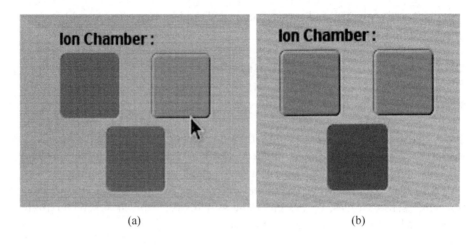

(a)   (b)

FIG. 19. *Different selections of AEC ionization chambers: (a) left and central ionization chambers selected; (b) central ionization chamber selected.*

2.2.17.4. *Analysis and interpretation*

Calculate the mean mAs value and exposure index for all the measurements. Compare each measured mAs and exposure index value with the corresponding mean values.

2.2.17.5. *Baselines and tolerances*

The recorded mAs and exposure index values should be within ±30% of the baseline values and within ±20% of the mean values.

Some specialized systems may use AEC sensors with different calibration factors, which affects the acceptability criteria; therefore, only the baseline values and AEC sensors with a similar calibration could be the basis for determining proper operation. Sensors that are relevant for the clinical practice should have a proper calibration.

2.2.17.6. *Frequency*

The test is repeated annually.

2.2.17.7. Corrective actions

Repeat the test if an outlying response is found. For this, reset completely the geometrical arrangements and the settings in order to exclude the possibility of errors, and make sure each time to select the appropriate ionization chamber.

After repeating the test, check whether the same sensor produces a value outside the tolerances. In this case, a recalibration may be necessary.

In addition, check the short term reproducibility (Section 2.2.18) if an error is suspected in the AEC system's controls.

If the repeated tests verify that there is a problem with the system, then request service support to adjust the AEC system. When the AEC needs recalibration, its use is restricted solely to manual exposure mode.

2.2.18. Automatic exposure control system — short term reproducibility

2.2.18.1. Description and objective

Similarly to the X ray generator, the AEC system should respond in a consistent manner. It is indispensable that under the same conditions, the AEC device determine the same exposure parameters in order to provide the lowest possible dose for the required image quality. This test is used to check the consistency of the AEC system's operation over successive exposures [4, 12].

2.2.18.2. Equipment

A suitable attenuator (e.g. PMMA slabs, a water filled tank, other homogeneous phantoms that cover the whole AEC sensors area) is needed for this test. The attenuator should have sufficient thickness to trigger an appropriate response from the AEC system (e.g. 15 cm). The arrangement is similar to the one in Section 2.2.17.

2.2.18.3. Procedure

The test procedure is as follows:

(a) Centre the tube on the image receptor, which is situated in the bucky. Set the SID at the grid focusing distance and ensure that the grid is in place.
(b) Place the attenuator on the table so that it covers the AEC sensors.
(c) Collimate the beam to cover the entire attenuator.
(d) Select the tube voltage used for this test previously (e.g. 70 kV) and the most frequently used AEC sensor(s).

(e)   Make an exposure and record the mAs value and exposure index obtained.

(f)   Repeat the previous step four more times. Use the same settings for all of these exposures.

2.2.18.4. Analysis and interpretation

Calculate the mean value of the mAs value and exposure index for all five measurements. Compare each mAs value and exposure index with the corresponding mean values.

2.2.18.5. Baselines and tolerances

The mAs value and exposure index should be within ±40% of the corresponding mean values.

2.2.18.6. Frequency

The test is repeated annually.

2.2.18.7. Corrective actions

If the tolerance is exceeded, then repeat the test, verifying that every arrangement corresponds to those used earlier (e.g. check whether the grid is in place). Request service support if the test fails again. When the AEC needs recalibration, its use is limited to manual exposure mode.

2.2.19. Automatic exposure control system — kilovoltage and thickness compensation

2.2.19.1. Description and objective

This test serves two purposes. First, it is used to check that the AEC produces a consistent response over a range of kilovoltage values. Second, it ensures that the AEC compensates for the thickness of the imaged object and provides a sufficient dose at the image receptor surface. The combination of these requirements necessitates that this test is performed on a regular basis [4, 12].

2.2.19.2. Equipment

A suitable attenuator (e.g. PMMA slabs, a water filled tank or other homogeneous phantoms that cover the whole AEC sensor area) is required for this test. For the thickness compensation test, the overall thickness of the attenuator varies; therefore, slabs of overall thickness 10, 15 and 20 cm should be used to limit the extent of the test to the thicknesses that are most frequently used clinically. Further investigation with thinner attenuators should be considered if paediatric patients or extremities are imaged regularly. For the arrangement, refer to Section 2.2.17.

2.2.19.3. Procedure

The test procedure is as follows:

(a) Kilovoltage compensation:
    (i) Centre the tube on the image receptor, which is situated in the bucky and set the SID at the grid focusing distance. Ensure that the grid is in place.
    (ii) Place the attenuator on the table so that it covers the AEC sensors. The chosen attenuator thickness should be an average of the available selection.
    (iii) Collimate the beam so that it covers the entire attenuator.
    (iv) Select a commonly used tube voltage (e.g. 70 kV) and the most frequently used AEC sensor(s).
    (v) Make an exposure with the selected settings and then record the mAs value and exposure index obtained.
    (vi) Repeat the previous step with two further settings, one higher and one lower, of the tube voltage (e.g. 60 kV and 85 kV).

(b) Thickness compensation:
    (i) Centre the tube on the image receptor, which is situated in the bucky. Set the SID at the grid focusing distance and ensure that the grid is in place.
    (ii) Place the attenuator on the table so that it covers the AEC sensors.
    (iii) Collimate the beam to cover the entire attenuator.
    (iv) Select a commonly used tube voltage (e.g. 70 kV) and the most frequently used AEC sensor(s). Retain this tube voltage throughout this part of the test.
    (v) Make an exposure and record the post-exposure mAs value and exposure index.
    (vi) Repeat the previous step with two further thicknesses of the attenuator, one larger and one smaller, as mentioned in Section 2.2.19.2.
    Note: For convenience, a combination of the kilovoltage values and attenuator thicknesses may be used, thus giving a total of nine

combinations instead of the three recommended above. This may help with recognizing a pattern in the operation of the AEC system.

2.2.19.4. Analysis and interpretation

The analysis and interpretation of the test results are carried out as follows:

(a) Kilovoltage compensation
     Compare the mAs value and exposure index with the corresponding baseline values and calculate their difference.
(b) Thickness compensation
     Calculate the mean mAs value and exposure index of all measurements performed with the different attenuator thicknesses. Compare every measured mAs value and exposure index with the corresponding mean values.

2.2.19.5. Baselines and tolerances

For each aspect of this test, the calculated difference should not be more than ±40% of the baseline value and the actual average.

2.2.19.6. Frequency

The test is repeated annually.

2.2.19.7. Corrective actions

If the tolerance is exceeded, then repeat the test, verifying that every arrangement corresponds to those used earlier (e.g. check whether the grid is in place). Significant deviations of the recorded values are not expected, even if minor changes occur. When the AEC does not compensate for the thickness or the kilovoltage as expected, its operability could be assessed with the short term reproducibility test, after checking the generator's overall performance (e.g. kilovoltage accuracy, linearity, reproducibility). Request service support if failure of either the AEC system or the generator is suspected.

2.2.20. Operation of the automatic exposure control guard timer

2.2.20.1. Description and objective

In case of system failure or technical error, the exposure needs to be terminated after a specific period of time (called 'backup time') or at a certain mAs value setting. This test ensures that the AEC device terminates the exposure at the proper setting of the guard timer [4, 12].

2.2.20.2. Equipment

The following equipment is needed for the test:

(a) An instrument capable of measuring exposure time in the range 5–5000 ms;
(b) A lead sheet with at least 2 mm thickness, large enough to cover the whole area of the AEC sensors.

2.2.20.3. Procedure

The test procedure is as follows:

(a) Centre the tube on the image receptor, which is situated in the bucky. Set the SID at a sufficient distance (e.g. 100 cm).
(b) Place the lead sheet in front of the beam.
(c) Arrange the instrument in the main axis of the X ray beam with its sensitive area facing the X ray tube (Fig. 20).
(d) Collimate the beam on the area where the AEC sensors are situated.
   Note: For convenience, the tube could also be directed away from the image receptor (e.g. aimed at the ground) and the collimator could be closed completely. Some X ray systems do not permit exposures under such conditions; thus, this set-up does not work universally.
(e) Select a low tube voltage ($\leq$60 kV) and set the guard timer at a reasonably low level (e.g. 40 mA · s). Select the most frequently used AEC sensor(s) for this test.
(f) Make an exposure with the above settings.
(g) Record the exposure time and take notes of any error messages that appear as feedback from this exposure.
   Note: Record the measured dose as well if the instrument has this capability.

FIG. 20. Set-up for the test of the AEC guard timer, with a lead plate placed beneath the solid state dosimeter.

2.2.20.4. Analysis and interpretation

Check that the AEC operation terminates at the set guard time or prior to reaching this setting.

2.2.20.5. Baselines and tolerances

The behaviour of the AEC system should be determined during commissioning. This should not change throughout the life cycle of the equipment. If changes occur, then consult the service engineer, as a change in the control software or a setting may affect how the AEC guard timer acts.

2.2.20.6. Frequency

The test is repeated annually.

2.2.20.7. Corrective actions

It is unlikely that the behaviour of the system would change. If that occurs, then check whether the AEC works properly. If it does not, request service support to investigate the cause of change. If the machine performs as expected otherwise, it could be used with due consideration.

2.2.21. Image uniformity and computed radiography plate sensitivity matching

2.2.21.1. Description and objective

This test is used to assess the degree and source of artefacts observed in digital radiography images and to confirm that the image of a homogeneous test object is uniform. It incorporates both visual and quantitative evaluation of image uniformity.

Since computed radiography systems use different image receptors, each plate should be tested individually. While this is a slightly different investigation from that used in digital radiography, the test could also be used to confirm the uniformity sensitivity of computed radiography plates and to determine the presence of plate related artefacts [4, 12, 16].

2.2.21.2. Equipment

A slab of PMMA with uniform thickness of approximately 20 cm or a 1 mm copper plate, preferably large enough to cover the entire image receptor (35 cm × 43 cm), is used for the test.

2.2.21.3. Procedure

The following procedure is used:

(a) Centre the X ray tube on the bucky using a distance of 100 cm from the focal spot to the image.
     Note: For computed radiography systems, a larger SID (at least 150 cm) and two exposures, one at 45° from the other, should be used to distinguish between intensity variations due to the heel effect and actual lack of uniformity.
(b) Collimate the beam to cover the entire image receptor area, and ensure that all AEC sensors are covered.

(c)   Place the PMMA slab in front of the bucky or slide the copper plate into the appropriate collimator slot. The measurement set-up is similar to those used for the AEC consistency tests (Section 2.2.17).

(d)   The exposure should preferably be carried out with the AEC. If the equipment does not have an AEC, then select the exposure factors manually (e.g. 70 kV and 3 mA · s for the copper plate). It is important to use the same exposure factors for computed radiography systems.
Note: For computed radiography systems, each time that a new plate is inserted into the bucky, record its identifiers and its condition (e.g. latches, hinges, integrity), along with the exposure index obtained from the readout.

(e)   Make the exposure and record the applied factors (e.g. AEC mode and settings, kilovoltage, mAs value).

(f)   Process the plate after a constant time delay (e.g. 30 s) to minimize the impact of latent image fading, using the same menu choices and image processing for each plate.

(g)   Record the exposure index and mAs value used for each cassette.

(h)   Repeat steps (d) through (g) for all plates.

2.2.21.4. Analysis and interpretation

The analysis and interpretation process is as follows:

(a)   Visually inspect each image on the workstation and inspect the image for artefacts (e.g. signs of scratches, scrapes, dents, dead pixels, patches) and uniformity. A narrow window width should be used for viewing the image. Record the presence or absence of significant artefacts.

(b)   Take five ROIs: one in the centre of the image and four in the centres of each quadrant. The ROI size should be about 100 pixels × 100 pixels.

(c)   Calculate the mean pixel number of the ROIs. Calculate the maximum difference of each ROI from the mean.
Note: If there is no linear relationship between pixel value and exposure, the mean values will need correction if the signal transfer properties of the image receptor are not linear.

(d)   Furthermore, in the case of computed radiography systems take the following steps:
    (i)   Calculate the mean exposure index and mAs value for all plates of the same size;
    (ii)   For each plate, determine the difference and the per cent difference between the mAs value and exposure index for that plate and the corresponding mean values.

2.2.21.5. Baselines and tolerances

The corrected ROI values should be within ±20% of the mean. The image should be without artefacts that could impair clinical use (e.g. dust, interference (moiré) lines, pixel failures, visible gridlines, distortion).

The manufacturer usually gives the tolerance for the exposure index of the computed radiography system. Typically, the mAs value used for a particular plate should be within ±5% of the mean value for plates of the same size.

2.2.21.6. Frequency

The test is repeated annually. For computed radiography systems, it is recommended to perform this test more frequently — every six months and after any service of the computed radiography reader that might affect its efficiency.

2.2.21.7. Corrective actions

If the tolerance is exceeded in a digital radiography system, then perform a flat field correction according to the manufacturer's method, or request help from the service engineer to carry out this test.

In the case of computed radiography plates, first try to clean them and check whether the reader's optics have been undergoing regular maintenance. Then repeat the test. Remove or replace the plates if the issue remains.

If the problem persists and the service engineer cannot perform a correction (e.g. for a dead pixel) or calibration, then record the position of the artefact and discuss it with the radiologists to evaluate its impact on diagnostic imaging.

Plates that do not perform within acceptable tolerances should be removed from clinical use. Plates with significant artefacts that cannot be erased by cleaning should be replaced. Computed radiography plates have a limited life expectancy and should be replaced if their sensitivity is significantly degraded. Inhomogeneity of the image should not affect clinical imaging and evaluation; if this occurs, then the imaging plate should be replaced.

2.2.22. Erasure efficiency (computed radiography only)

2.2.22.1. Description and objective

Computed radiography imaging plates have to be read out and erased in order to be reusable for imaging. Owing to the technology used to produce images by computed radiography systems, residual signals may be present on subsequent images. Erasure should be thorough to prevent ghosting (the reappearance of latent images) and needs to be tested on a regular basis to evaluate its performance [16].

2.2.22.2. Equipment

The following equipment is needed for the test:

(a)   A piece of high attenuation material with a size smaller than the imaging plate to be used (e.g. a 2 mm thick lead sheet, or a copper plate more than 3 mm thick and with an area between 5 cm^2 and 15 cm^2);
(b)   A lead apron.

2.2.22.3. Procedure

The test procedure is as follows:

(a)   Place the lead apron on the table.
(b)   Place the freshly erased computed radiography plate on the lead apron and centre the X ray tube on the plate at a sufficiently large distance of 150–180 cm.
(c)   Adjust the collimators to expose the whole area of the computed radiography plate and position the attenuator in the centre (Fig. 21).
(d)   Select a voltage of 60 kV and an appropriate mAs value so as to expose the computed radiography plate to approximately 50 µGy in manual exposure mode. No additional filtration is required.
(e)   Process the computed radiography plate by applying the default imaging procedure recommended by the manufacturer. Save this image, and then prepare the computed radiography plate for the next exposure and erase it.
(f)   Place the same computed radiography plate on the lead apron, at the same position as for the first exposure.
(g)   Expose the computed radiography plate again, using a fraction of the mAs value used previously. Set an appropriate mAs value so as to expose the plate to approximately 1 µGy but, during this exposure, without any attenuator.
(h)   Read the computed radiography plate with the same parameters.
      Note: A quantitative analysis is described in detail in Ref. [16], involving further processing of the computed radiography plate.

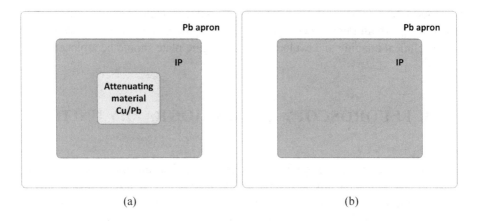

(a)   (b)

FIG. 21. Set-up for testing the erasure efficiency of computed radiography imaging plates. (a) Exposure of computed radiography plate with attenuating material; (b) exposure of the computed radiography plate without attenuating material.

2.2.22.4. Analysis and interpretation

Visually inspect the image for any trace of the first image. Evaluate the images by viewing them on a display used in clinical practice to diagnose patients, with the appropriate levelling and window width.

2.2.22.5. Baselines and tolerances

Absence of a ghost image of the attenuating material in the second image demonstrates acceptable erasure thoroughness. In any case, there should be no trace of the ghost image, as this could impact clinical use of the imaging system.

2.2.22.6. Frequency

The test is repeated annually.

2.2.22.7. Corrective actions

Erase the computed radiography plate and repeat the test if ghosting is present. Some computed radiography reader systems have adjustable erasure time, and recalibration of this parameter may be sufficient to eliminate ghosting. A persistent image may require the replacement of the light source used for erasure and may indicate that the computed radiography plates are deteriorating with extensive use, so request service support to investigate the cause of ghosting. Inhomogeneity of the image should not affect clinical imaging and evaluation. If inhomogeneity affects clinical imaging and evaluation, the imaging plate should be replaced.

3. FLUOROSCOPY AND ANGIOGRAPHY UNITS

For fixed radiography and fluoroscopy units, all applicable tests in Section 2 should be performed, since many of these X ray units operate also in radiographic imaging mode. For mobile units (e.g. C arms), relevant testing of the tube parameters should be performed in fluoroscopy mode with AERC, also called automatic brightness control or automatic dose rate control selection. For tests of the radiation generator, remarks can be found in Section 2, as AERC requires an attenuator to operate properly. For the fluoroscopy mode of operation, additional filtration can vary considerably with system settings and imaging protocol. Therefore, it is particularly important to use the same imaging settings for the tests that are used clinically.

3.1. QUALITY CONTROL TESTS FOR RADIOGRAPHERS

3.1.1. Reproducibility of the automatic exposure rate control

3.1.1.1. Description and objective

AERC is mandatory for all fluoroscopy units. AERC adjusts the exposure parameters automatically in order to ensure a predefined image quality under the constantly changing attenuation conditions of the X ray beam. The constancy of the AERC settings with different loadings, phantoms and settings should be regularly tested to ensure that the AERC system is operating appropriately [7].

3.1.1.2. Equipment

The following equipment is required for the test:

(a) PMMA slabs that can cover the whole area of the image receptor, with thickness of 10–20 cm. Copper plates of 1 and 2 mm thickness that cover the whole area of the image receptor can be used as an alternative.
(b) Measuring tape or ruler.

3.1.1.3. Procedure

The test procedure is as follows:

(a) Set the focal spot to image receptor distance to be 100 cm or another clinically relevant distance used earlier, and set the arm holding the X ray tube and the image receptor to be vertical.
    Note: In many cases, it is more convenient to rotate the X ray tube into an upside down arrangement, with the X ray tube on the upper end of the C arm, rather than in the regular 'under couch' set-up.
(b) Position the table or couch near the image receptor.
(c) Place 10 cm of PMMA on the couch or the table, or place a copper plate with 1 mm thickness either on the collimator or the table (Fig. 22).
(d) Note or draw the exact geometry of this arrangement and measure the focal spot–image receptor distance, the focal spot–table distance and the focal spot–test object distance.

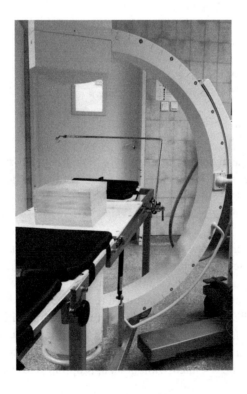

FIG. 22. Simple set-up using PMMA to check the performance of AERC settings.

(e)   Set the most frequently used imaging protocol (i.e. additional filtration), focal spot size and mode of operation (e.g. high, medium or low dose rate mode, pulse rate mode). The field of view (FOV) should also be the largest possible. Ensure that the grid is in the same arrangement as when determining the baseline. Note these settings.

(f)   Make an exposure long enough for the AERC to reach a constant value (e.g. for more than 5 s) and note the kilovoltage, current, selected additional filters, pulse width and frames per second. Check that the set parameters of the previous step are correct. Record every relevant feedback value.

(g)   Change the selected field size and repeat steps (d) through (f) for each clinically relevant magnification or FOV.

(h)   Repeat steps (c) to (g) with different thicknesses of the attenuating objects (e.g. 15 cm and 20 cm PMMA, 2 mm and (2 + 1) mm copper plate).

3.1.1.4.   Analysis and interpretation

For each setting and parameter, compare the results of the readings with the previously established baseline.

3.1.1.5.   Baselines and tolerances

The tolerance with respect to the baseline for the kilovoltage is ±5%; for the current, it is ±20%.

3.1.1.6.   Frequency

The test is repeated every three months. The frequency of the test should be adjusted according to the results and the decision of the medical physicist.

3.1.1.7.   Corrective actions

Whenever a difference outside the indicated tolerances is observed, check the exposure parameters and the geometry of the arrangement. If the arrangement is the same as that used for the establishment of the baseline values, then repeat the test after completely resetting the measurement and the set-up.

Ask the medical physicist to perform a more thorough investigation if the deviation persists, as this could severely impact image quality. Eventually, if the error is reproducible, the service engineer should be contacted. The X ray machine should not be operated without an appropriately working AERC unit.

3.2. QUALITY CONTROL TESTS FOR MEDICAL PHYSICISTS

3.2.1. Verification of beam collimation

3.2.1.1. Description and objective

Verify that the X ray beam is collimated so that the total exposed area remains within the edges of the image receptor for the largest FOV [14].

3.2.1.2. Equipment

The following equipment is used:

(a) Phantom for checking the collimation of the field, containing a grid and marks with known dimensions, or a fluorescent screen with marks;
(b) Computed radiography plate or self-developing film;
(c) Measuring tape or ruler.

3.2.1.3. Procedure

The following procedures are used for the test:

(a) Measuring the field size with an imaging device:
 (i) Set the arm holding the X ray tube and the image receptor to be vertical. **Note:** In some cases, an upside down arrangement of the X ray tube and image receptor may be useful.
 (ii) Place the computed radiography plate or a similar imaging device on the image receptor of the fluoroscopy unit or as close to it as possible.
 (iii) Place the phantom on the imaging device. Note the distance of each object in this arrangement and calculate the magnification factor.
 (iv) Move the X ray tube to the closest possible position to the image receptor and choose the largest possible field size by fully opening the beam limiting device.
 (v) Expose the phantom to an appropriate load to acquire an image, and note the exposure parameters.
 (vi) Acquire the image from the imaging device.
 (vii) Repeat steps (ii) through (vi), with the imaging device at the farthermost position from the image receptor.
(b) Measuring the field size with a fluorescent screen:
 (i) Move the X ray tube to the closest possible position to the image receptor and choose the largest possible FOV.

(ii) Dim the ambient lights in the room.

(iii) While wearing a lead apron, move the fluorescent screen around the edges of the image receptor during an exposure, and evaluate whether the beam irradiates areas beyond the edges of the image receptor of the fluoroscopy unit.

(iv) Note whether the X ray beam is not collimated to the edges of the image receptor and measure its extent (Fig. 23).

3.2.1.4. Analysis and interpretation

The images are evaluated by measuring the distance between the edge of the X ray field and the nominal edge of the image receptor at the plane of the imaging device or the fluorescent screen. The measurements should be corrected for the magnification factor. Refer to Sections 2.1.1 and 2.2.1.

3.2.1.5. Baselines and tolerances

Regardless of the method, the X ray field should not deviate from the image receptor area by more than 2% of the distance between the focal spot and the image receptor.

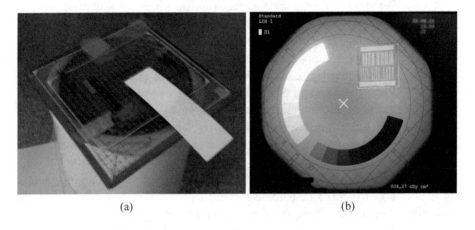

(a)   (b)

FIG. 23. (a) Set-up used to verify the collimation of the X ray beam using a fluorescent screen. (b) The iris collimator is visible on the edge of the image, indicating that the beam confinement is acceptable.

The test is repeated annually.

Repeat the test after completely resetting the set-up. Check for visible mechanical defects at the X ray tube assembly or at the collimator. Verify whether the beam limiting device is faulty at all selectable field sizes (Section 3.3.8). Depending on the situation, it may be necessary to limit or suspend work until the beam limiting device is repaired.

3.2.2. Verification of beam geometry

Most fluoroscopy units do not have a light field indicator. If the system is equipped with one, then perform the test described in Section 2.1.1 (adjusted as needed for the type of equipment, such as fixed unit or C arm) [4, 5], along with the test in Section 3.3.6. These tests should be performed annually by the supervising medical physicist.

3.2.3. Verification of different field sizes

3.2.3.1. Description and objective

Magnification and proper collimation play an important role in limiting the patient's exposure. This test is useful for evaluating the field size and ensuring that the collimator is working properly. It is used to validate the conformity of the nominal and the automatically set FOVs [14].

3.2.3.2. Equipment

The equipment required is a phantom (with marks or an appropriately sized mesh with known dimensions) larger than the whole image receptor area. It is not convenient to use coins or other similar markers for this test.

3.2.3.3. Procedure

The following procedure is used:

(a) Set the arm holding the X ray tube and the image receptor to be vertical.
(b) Place and centre the test object on the image receptor.

(c)   Choose the largest FOV and magnification and open the collimator fully.

(d)   Choose the next FOV size.

(e)   Expose the phantom and let the AERC set the parameters to acquire an image; note these parameters while evaluating the image.

(f)   Choose a smaller FOV and repeat steps (d) and (e) until all FOVs are checked (Fig. 24).

3.2.3.4. *Analysis and interpretation*

The following process is used to analyse and interpret the test results:

(a)   Evaluate the displayed field size for every FOV.

(b)   Determine whether there is a deviation from the nominal FOV.

3.2.3.5. *Baselines and tolerances*

The X ray field size should not deviate from the nominal area by more than ±2% of the SID.

3.2.3.6. *Frequency*

The test is repeated annually.

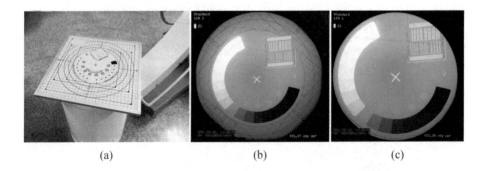

(a)   (b)   (c)

FIG. 24. (a) Set-up used to verify the field size of a C arm fluoroscopy unit using a test object. Images with a field of view diameter of (b) 23 cm and (c) 16 cm.

3.2.3.7. Corrective actions

Repeat the test after completely resetting the arrangement. Check for visible mechanical defects at the X ray tube assembly or at the collimator. Ask the manufacturer's service engineer to adjust the beam limiting device.

3.2.4. Patient entrance surface air kerma rate

3.2.4.1. Description and objective

The aim of this test is to verify that the maximum achievable entrance surface air kerma (ESAK) rate is under the institutional or national limit. In addition, the measured results are useful for the evaluation of expectable peak skin dose levels [6, 9, 11]. It is important to emphasize that monitoring of the peak skin dose during a given procedure and assessment of patient exposure should be complementary to this test.

3.2.4.2. Equipment

The following equipment is required for the test:

(a) A dosimeter with a calibrated ionization chamber or a solid state detector. The detector should not impact the AERC settings.
   Note: Solid state dosimeters that contain lead backscatter shielding should not be used, unless they are specifically designed for the purpose of measuring dose rate on fluoroscopic units, as the lead insert in the instrument may impact the AERC significantly, resulting in an exposure higher than the set value.
(b) A water phantom 20 cm thick, covering the image receptor. Alternatively, an 18.5 cm thick PMMA plate set may be used, but in this case, corrections should be introduced according to Ref. [9] (appendix VII and VIII).

3.2.4.3. Procedure

The arrangement for this measurement is illustrated in Fig. 25. The procedure is as follows:

(a) If an antiscatter grid is in use clinically, make sure it is in position.
(b) Position the phantom on the couch.
   Note: To simulate a larger patient, an additional 10 cm of water or equivalent PMMA may be added.

(c) Position the measuring instrument in contact with the phantom and in the centre of its entrance surface, facing the X ray beam.

(d) Set the distance between the focal spot and the image receptor to that used in clinical practice. There are no standard values for this distance; however, 100 cm or the nearest possible distance to 100 cm could be used, similarly to the other tests.

(e) Open the collimators to the size of the image receptor.

(f) Measure and record the focal spot–image receptor and focal spot–radiation detector distances.

(g) Expose the phantom using a clinical imaging protocol and record the dosimeter reading, kilovoltage and current, as well as the image receptor settings; then repeat this exposure two more times.

(h) Repeat step (g) for all FOVs, dose rate settings and different imaging protocols used under normal clinical conditions.

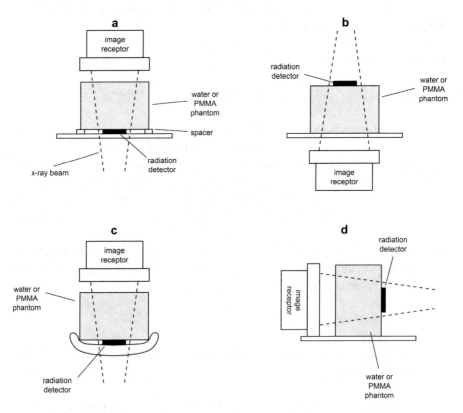

FIG. 25. Configuration for the measurement of patient entrance surface air kerma rate: (a) under couch installation; (b) over couch installation; (c) C arm unit with ancillary couch; (d) C arm unit set for lateral exposures or when a couch is not available for clinical use.

3.2.4.4. Analysis and interpretation

Apply the appropriate correction factors (e.g. pressure, temperature, calibration coefficient) for the instrument and correct the measured ESAK rates to the reference point using the inverse square law. When performing the evaluation, keep in mind that the ESAK includes backscatter, while incident air kerma does not.

In the case of an angiography machine, the patient entrance reference point is set by the manufacturer, and it mainly depends on the type of the interventional system, according to the IEC 60601-2-54 standard [11]:

— When the tube is mounted below the patient support, the patient entrance reference point is 1 cm above the table.
— When the tube is above the patient support, the patient entrance reference point is located 30 cm above the patient support.
— For C arm units, the patient entrance reference point is 15 cm from the isocentre in the direction of the focal spot (e.g. 35 cm from the focal spot if the isocentre is defined and the focal spot–image receptor distance is 1 m).

3.2.4.5. Baselines and tolerances

The manufacturer of the equipment often provides a reference statement on incident air kerma rate, which may be used as a basis for comparison. This is applicable only to arrangements with no backscatter at all; therefore, measurements taken according to the above method should be corrected for backscatter using the factors provided in Ref. [8].

National regulations specify certain values for the ESAK and need to be taken into consideration. If there are no national regulations, then values from various international standards or guidelines can be used. For example, a maximum value of 100 mGy/min for normal mode can be applied according to Refs [5, 6]. According to standard IEC 60601-2-54, the incident air kerma rate should be less than 88 mGy/min for any setting of the normal or low dose rate mode of operation, and 176 mGy/min is permitted for the high dose rate or high level control mode [11].

3.2.4.6. Frequency

The test is repeated annually.

Check the exposure parameters, the calculations and the arrangements of the set-up. Repeat the test, completely resetting the geometrical arrangement. Depending on the observed deviations, it may be necessary to limit or suspend work with the machine until it is repaired. Perform a thorough investigation on the possible sources of error or deviation of the results.

3.2.5. **Image receptor entrance surface air kerma rate**

3.2.5.1. *Description and objective*

This test monitors the constancy of the imaging chain. This test is used to verify that the air kerma rate at the entrance of the detector conforms to the maximum allowed level [4, 7].

3.2.5.2. *Equipment*

The following equipment is required for the test:

(a) A dosimeter with a calibrated ionization chamber or solid state detector;
(b) An aluminium attenuator with a thickness of 25 mm and copper plates, as necessary.
     Note: Alternatively, a 20 cm thick water or equivalent phantom that covers the image receptor can be used.

3.2.5.3. *Procedure*

The arrangement for this measurement is similar to that described earlier (see Fig. 8), but the instrument is placed on the image receptor. Figure 26 shows a possible arrangement with the X ray tube at the top.
The test procedure is as follows:

(a) If an antiscatter grid is in use clinically, make sure it is in position.
(b) Position the phantom on the couch.
     Note: To simulate a larger patient size, 10 cm of water or equivalent PMMA may be added.
(c) Position the instrument in contact with the image receptor, as close as possible to its entrance surface and facing the X ray beam. If further attenuating layers are present (e.g. bucky, table, AERC sensors), then record these during the measurement.

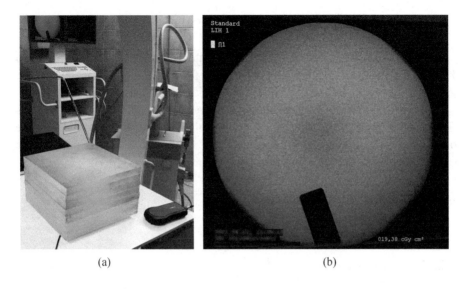

(a)	(b)

FIG. 26. (a) Set-up used to test the air kerma rate at the entrance of the detector, showing the solid state detector positioned below the PMMA plates. (b) The obtained X ray image.

(d) Set the focal spot–image receptor distance to that used in clinical practice. There are no standard values for this distance; however, 100 cm or the nearest possible distance to 100 cm could be used, as in the other tests.

(e) Open the collimators to the size of the image receptor.

(f) Measure and record the focal spot–image receptor and focal spot–radiation detector distances.

(g) Expose the phantom using a clinical imaging protocol and record the instrument's reading, kilovoltage and current, as well as the image receptor settings; then, repeat this exposure two more times.
Note: The IEC recommends using additional copper attenuators to enable the AERC to set a tube voltage between 70 and 80 kV.

(h) Repeat the previous step for all field sizes, dose rate settings and different imaging protocols used under normal clinical conditions.

3.2.5.4. Analysis and interpretation

Apply the appropriate correction factors for the instrument and correct the measured air kerma rates to the entrance surface of the detector using the inverse square law. Correct for the attenuating layers present between the instrument and the image receptor. The transmission factors are usually found in the documentation provided by the manufacturer.

3.2.5.5. Baselines and tolerances

In fluoroscopy mode, the air kerma rate is typically between 0.2 and 1 μGy/s if an aluminium attenuator is used and filters are added to achieve a tube voltage between 70 and 80 kV. This value applies for grid attenuation.

The difference of the measured results from the established baseline values should not differ by more than ±25% ('remedial level') according to the Institute of Physics and Engineering in Medicine [4]. Otherwise ±50% is used as the suspension level.

3.2.5.6. Frequency

The test is repeated annually.

3.2.5.7. Corrective actions

Check the exposure parameters, the calculations, and the arrangements of the set-up. Repeat the test, completely resetting the geometrical arrangement. Whenever an issue is encountered with the AERC, it is advisable that the system is recalibrated by a service engineer. Until recalibration or repair, depending on the observed deviations, it may be necessary to limit or entirely suspend the use of the X ray unit.

3.2.6. Kerma–area product meter calibration

The test is as described in Section 2.2.7.

3.2.7. Leakage radiation

The test is as described in Section 2.2.11.

3.2.8. Scattered radiation

The test is as described in Section 2.2.12.

3.2.9. Image quality of fluoroscopy

3.2.9.1. Description and objective

The image quality assessment for fluoroscopy consists of testing several indicators related to image quality. Descriptions of the recommended tests are provided in Section 2 (not repeated here for simplicity) and do not need any further adaptation to apply to fluoroscopy. The recommended tests for assessing the image quality of fluoroscopy systems include the following:

(a) Low contrast detectability test (see Section 2.2.13);
(b) Spatial resolution test (see Section 2.2.14);
(c) Image uniformity test (see Section 2.2.21).

The image uniformity test normally involves only the evaluation of a homogeneous test image, while in the case of fluoroscopy, its aim is to confirm that the image is not distorted and that it is free of artefacts.

By repeating the test using different attenuators to simulate the patient and the clinically applicable scatter conditions, one may also check that the AERC system is compensating the attenuating properties of the imaged object (see Section 2.2.19).

In the following, a combined test is described. It should be noted that this is a very simplified test, and if images of the test objects can be observed in various phantoms, then the exposures should be repeated [4, 15].

3.2.9.2. Equipment

The image quality phantom recommended for radiography is suitable for fluoroscopy. However, a combined image quality phantom simplifies the test because it includes the following:

(a) Low contrast inserts;
(b) Mesh, grid or another test object to check geometrical distortion (especially relevant for image intensifiers);
(c) Spatial resolution test pattern;
(d) High contrast (copper) step wedge (to check the image display).

Figure 27 shows such a phantom.

To carry out this test, it is recommended to use PMMA plates or a water phantom (covering the image receptor area and having a total thickness of 20 cm) or as recommended by the manufacturer of the test object.

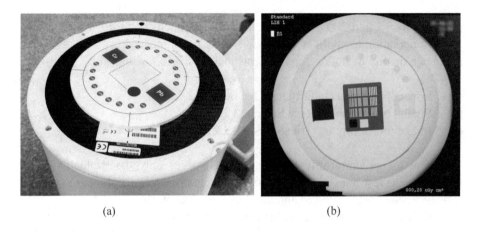

(a)   (b)

*FIG. 27. Set-up used to test image quality in fluoroscopy using an image quality phantom.
(a) Fluoroscopy image quality phantom; and (b) the corresponding X ray image.*

3.2.9.3. Procedure

The test procedure is as follows:

(a) If an antiscatter grid is in use clinically, make sure that it is in position.

(b) Position the combined test object on the image receptor or as close to it as possible.

(c) Set the focal spot–image receptor distance equal to that used in clinical practice. There are no standard values for this distance; however, 100 cm or the nearest possible distance to 100 cm could be used, similarly to the other similar tests.

(d) Open the collimators to the size of the image receptor.

(e) Measure and record the focal spot–image receptor distance.

(f) Expose the test object using a clinical imaging protocol and record the dosimeter reading, kilovoltage and current, as well as the image receptor settings.

(g) Repeat the previous step for all field sizes, dose rate settings and different imaging protocols used under normal clinical conditions.

(h) Optionally, repeat step (g) by placing 10, 15 and 20 cm thick PMMA plates or other suitable attenuators in front of the image quality phantom.

3.2.9.4. Analysis and interpretation

The acquired images are evaluated on the image display used in clinical practice. The high contrast step wedge insert of the image quality phantom is used to check the brightness settings of the image display. This is evaluated without any additional attenuator.

The test conditions and results are evaluated after each exposure.

3.2.9.5. Baselines and tolerances

Evaluate whether the low contrast sensitivity has changed over time and compare the results with the baseline values. The low contrast resolution should not change significantly. Since there are no standard test objects, no further numerical values can be given.

There should be no artefacts on the image, but parallel lines of the mesh grid may appear.

If the test is carried out with different scattering objects and the change of the imaging parameters is documented, then these should be compared with the baseline values. It is expected that under the same conditions, the kilovoltage, current, added filters and other imaging parameters do not change significantly. The way that the exposure rate and the radiation quality are determined by the AERC should not change significantly when using a consistent imaging protocol.

3.2.9.6. Frequency

The test is repeated annually.

3.2.9.7. Corrective actions

When an error is suspected, check the results and the imaging parameters and then repeat the test. If the low contrast visibility is worse than expected or artefacts are present, request service support.

If inconsistent X ray parameters set by the AERC are encountered, then the AERC system may need to be checked by a service engineer.

Whenever major changes occur to the X ray generator, the X ray tube assembly or the software, then a new baseline may need to be set. Consult with the users to determine whether the change in image quality will affect clinical use. Depending on the situation, it may be necessary to limit or suspend work with the machine until it is repaired.

3.2.10. Image quality of digital subtraction angiography

3.2.10.1. Description and objective

The aim of the test is to check the constancy of the image quality parameters in the digital subtraction angiography modality. This mode of operation is not affected in modern systems with digital image receptors, but this test has an important role in systems using image intensifiers. This requires a special phantom and the appropriate timing to change the position of the moving insert of the phantom [17].

3.2.10.2. Equipment

The equipment required for the test is a phantom specially designed to measure the dynamic range, contrast sensitivity, artefacts and non-linearity compensation of the system (see Figs 28, 29).

3.2.10.3. Procedure

Perform image acquisition according to the instructions of the phantom's manufacturer and record the relevant image acquisition parameters.

3.2.10.4. Analysis and interpretation

The acquired image should be evaluated according to the instructions of the phantom's manufacturer.

3.2.10.5. Baselines and tolerances

Evaluate whether the contrast sensitivity has changed over time and compare the results with the baseline values. The image should have no visible artefacts and no logarithmic errors (applicable only to systems equipped with image intensifiers).

3.2.10.6. Frequency

The test is repeated annually.

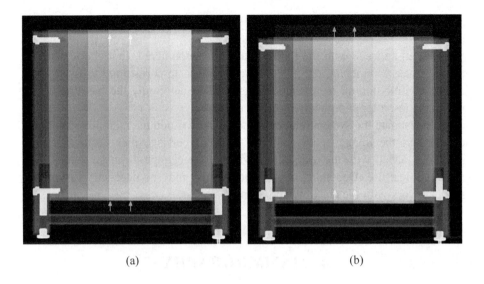

(a)   (b)

FIG. 28. Image of the phantom used for testing the image quality of a digital subtraction angiography system: (a) in the original position; and (b) in the subtraction position.

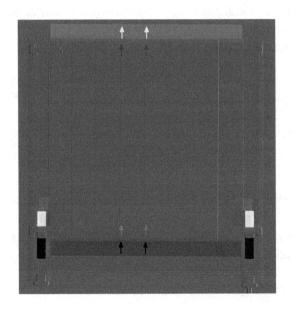

FIG. 29. Image of the phantom in Fig. 28 after subtraction.

When an issue with the image quality is encountered, first check whether the exposure parameters, the phantom and the set-up correspond to those used during commissioning or to the settings used to determine the baseline. Then repeat the test after completely resetting the set-up.

Investigate the system (e.g. evaluate the AERC and the radiation generator, then assess the image quality). If the problem persists, then request service support. Consult with the users about whether the change in image quality will affect clinical use. Depending on the situation, it may be necessary to limit or suspend work with the machine until it is repaired.

4. MAMMOGRAPHY

Mammography is a modality specialized for imaging of the breast. New radiation qualities are available for breast imaging in addition to molybdenum target–molybdenum filter systems, which are still used as a benchmark for image quality and radiation dose. The latest commercially available modalities are breast tomosynthesis and contrast enhanced mammography. However, QC guidance for such systems did not develop as quickly as technology evolved. As there are no widely accepted consensus documents available to check new technology, this section is limited to describing the 'traditional' projection imaging QC tests.

As in other sections of this handbook, these test descriptions are limited to the essentials, not requiring expensive equipment or sophisticated software. For this reason, some tests are omitted which are considered important by other publications, such as low contrast detectability.

4.1. QUALITY CONTROL TESTS FOR RADIOGRAPHERS

4.1.1. Image receptor uniformity — weekly test (computed radiography and digital radiography only)

4.1.1.1. Description and objective

Uniformity should be visually checked on a weekly basis using the image of 45 mm thick rectangular PMMA slabs. By setting a narrow grey scale window, the viewer can see areas of non-uniformity in order to identify artefacts that might disturb or compromise diagnosis [18].

4.1.1.2. Equipment

The equipment required is uniform rectangular slabs of 45 mm thick PMMA or a phantom provided by the manufacturer. The phantom should cover the whole detector area and have no deep scratches in order to be sufficiently uniform.

4.1.1.3. Procedure

The test procedure is as follows:

(a) Place the uniform PMMA slabs on the breast support, taking care to completely cover the detector.
(b) Mount the compression paddle.
(c) Ensure that the acquisition workstation is set to save images 'for processing' and that any possible image processing algorithm is disabled.
(d) Acquire an image in the automatic exposure mode that is most often used clinically, with the most frequently used AEC sensor, if it is selectable.
(e) Record the exposure settings (i.e. kilovoltage, target/filter, density control setting (if applicable), AEC mode and post-exposure mAs value).
(f) For computed radiography systems, the search for artefacts should be performed for each imaging plate in clinical use.

4.1.1.4. Analysis and interpretation

The analysis and interpretation of the test results are carried out as follows:

(a) Evaluate the unprocessed image by measuring the MPV and standard deviation in ROIs (with an area of 1 cm^2) placed at five specific positions on the image: one centrally located and the other four near the corners of the image, as shown in Fig. 30.
(b) Record the MPV in the central ROI (MPV$_c$).
(c) Record the MPV in the other ROIs (MPV$_e$).
(d) Check the homogeneity visually. The window width should be set at 10% of the MPVs.
(e) Identify the MPV$_e$ value with the largest difference from MPV$_c$.

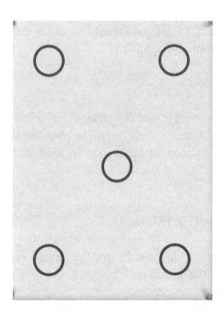

FIG. 30. *Image used to test receptor uniformity. Shown in red are the positions of the regions of interest used for the measurement of the mean pixel value and the standard deviation.*

(f)   Calculate the maximum deviation (in per cent) of the MPV_e value found in step (e) from the central value according to Eq. (4):

$$\text{maximum deviation} = \frac{MPV_c - MPV_e}{MPV_c} \times 100 \qquad (4)$$

4.1.1.5.   Baselines and tolerances

The maximum deviation in MPV should be lower than ±15% of the MPV in the whole image.

4.1.1.6.   Frequency

The test is repeated weekly.

4.1.1.7.   Corrective actions

Check the exposure parameters and geometry set-up, and then repeat the test, completely resetting the measurement and exposure set-up.

If there are artefacts, rotate the phantom to check the source of artefact. Contact the medical physicist to perform a more thorough investigation if artefacts are again found on the image(s). In principle, artefacts are not acceptable. However, as previously remarked, the urgency of remedial actions depends on the nature of the artefact. In general, subtle artefacts mimicking clinical features (potentially producing unnecessary biopsies) could require suspending use of the X ray system or a given computed radiography plate.

4.1.2. Subjective image quality evaluation

4.1.2.1. Description and objective

Evaluation of the image quality from images obtained by exposing test objects or phantoms provides a practical approach to characterizing image quality performance during commissioning of a new piece of equipment and in routine assessments of equipment performance over time [18].

4.1.2.2. Equipment

The following equipment is required:

(a) Nationally or internationally recommended phantom used by the facility for its QC programme for film–screen or digital mammography systems. It is recommended to use a breast mimicking phantom (see Fig. 31), which preferably has inset objects similar to the characteristics of a breast (e.g. microcalcifications, tumour-like masses, fibrous structures similar to the speculums).
(b) Phantom image from the baseline evaluation.

4.1.2.3. Procedure

The test procedure is as follows:

(a) Align the phantom on the breast support with the chest wall and centre it.
(b) Lower the compression paddle to apply a minimal compression force.
(c) Select exposure parameters that would be used clinically for a similarly thick breast. Also set the appropriate target, filter, kilovoltage, grid, AEC sensor and operation mode (semiautomatic or automatic).
(d) Record the exposure parameters, including the exposure index in the case of computed radiography systems.

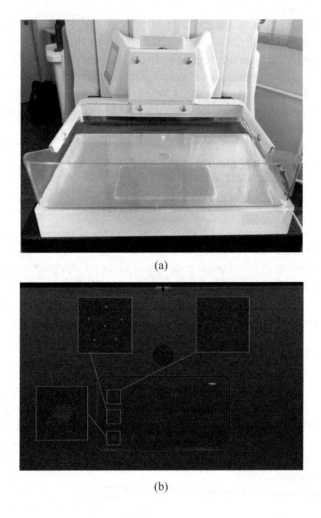

FIG. 31. Set-up used to test image quality using an image quality phantom. (a) Measurement arrangement and (b) X ray image of a phantom containing insets mimicking breast tissue.

(e) Process the image using the algorithms that would be used clinically.

(f) For computed radiography systems, this test should be performed for each imaging plate in clinical use.

4.1.2.4. Analysis and interpretation

Compare the image with the one made at the time of acceptance, which is used as a baseline. Evaluate the image for artefacts and determine whether these would impact the image. Use the magnifying lens to search for signs of inhomogeneities (i.e. dirt, dust or lines if the grid was used) and check the image for artefacts.

Use the evaluation method recommended by the manufacturer of the test object. Record the results and investigate the cause if problems are observed.

4.1.2.5. Baselines and tolerances

The following requirements apply:

(a) The mAs value used for the phantom exposure should be within ±10% of the baseline mAs value for the same kilovoltage and filter.
(b) There should be no significant degradation of image quality from the baseline image.
(c) There should be no blotches or regions of altered noise appearance.
(d) There should be no observable lines or structural artefacts.
(e) There should be no 'bright' or 'dark' pixels evident.

4.1.2.6. Frequency

The test is repeated weekly.

4.1.2.7. Corrective actions

If the results are outside the tolerances or are unacceptable, the tests should be repeated. The medical physicist should check the following:

(a) When image quality deteriorates over time, then the source of the change should be investigated. The parameters used for imaging (e.g. kilovoltage, AEC, display, processing algorithms) should be reviewed to investigate changes in these parameters.
(b) A software upgrade may change the AEC settings. This needs to be examined and, if confirmed, the service engineer needs to be contacted.
(c) Artefacts that mimic tissues can severely impact diagnosis; if such artefacts are observed, then flat fielding of the image receptor is necessary. Repeat the test after this calibration.
(d) The authorized service engineer may provide further help.

4.2. QUALITY CONTROL TESTS FOR MEDICAL PHYSICISTS

4.2.1. Compression force and thickness indicator accuracy

4.2.1.1. Description and objective

The objective of this test is to confirm that the mammography system provides an adequate compression in manual and automatic mode, and to check whether the compression plate tilts. The accuracy of the compression force indicator, if present on the equipment, can be tested as well. The accuracy of the compression thickness indicator is also checked during this test, and it may have a significant role in several imaging systems in which the AEC mode determines the radiation quality to be used for the exposure according to the thickness of the breast [18].

4.2.1.2. Equipment

The following equipment is required for the test:

(a) Analogue type bathroom scales or other device capable of measuring force;
(b) Bath towels or blocks of rubber foam;
(c) Slabs of PMMA used for AEC testing (20, 45 and 60 mm thick; preferred size: 18 cm × 24 cm for non-digital systems and 24 cm × 30 cm for digital systems).

4.2.1.3. Procedure

The test procedures are as follows:

(a) Automatic compression:
  (i) Place a towel on the surface of the breast support to avoid scratches.
  (ii) Place the scale on the towel and centre it below the compression paddle (see Fig. 32).
  (iii) Place towels or rubber foam on the scales to protect the compression paddle.
  (iv) Set the maximum permitted automatic compression force by using the pedal more than once.
  (v) Compare and record the readings of the indicated force from both the system and the scales.

(b) Manual compression:
(i)   Using the manual compression mode, move the compression paddle until it stops.
(ii) Record the compression force.
(iii) Release the compression.
      Note: The above measurements could be repeated at other force settings. Cases where the compression is lost without interaction need to be recorded.
(c) Compression thickness indicator:
(i)   Centre and align the PMMA slabs to the chest wall edge of the breast support.
      Note: Some mammographic systems are calibrated to account for the tilting of the compression paddle. Higher compression force will cause the indicated value of the compression force to rise because of tilting. When 18 cm × 24 cm PMMA slabs are used, tilting is possible.
(ii) Apply a compression force that is used clinically and record its value. The same force should be used for all consecutive measurements.
(iii) Record the thickness indicated on the display and the actual thickness of the slab.
(iv) When the AEC selects the exposure parameters according to the indicated thickness and a special table is used for the magnification mode, repeat this measurement.

4.2.1.4. Analysis and interpretation

The reading of the scales should be recorded in the same units as those used for display on the mammography machine, and corrected if necessary.

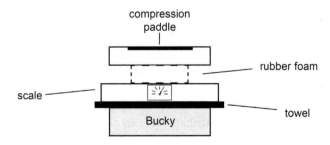

FIG. 32. Set-up used to perform compression force measurement with bathroom scales.

4.2.1.5.   Baselines and tolerances

The following baselines and tolerances apply:

(a)   The maximum compression force for automatic compression should be 150–200 N;
(b)   The maximum manual compression force is 300 N;
(c)   The displayed value accuracy should be ±20 N;
(d)   The displayed thickness should be within ±8 mm of the slab thickness.

4.2.1.6.   Frequency

The test is repeated annually or when there is an observed reduction in the breast compression quality.

4.2.1.6.   Corrective actions

If the measured values are outside the tolerance range or if compression is lost without interaction, the compression device should be calibrated by a service engineer. If safe and efficient breast compression is compromised, use of the system should be suspended.

4.2.2.   Detector alignment, alignment of X ray field to detector area

4.2.2.1.   Description and objective

This test may be used to determine the amount of breast tissue at the chest wall that is excluded from the image, which is affected by the imaging geometry or detector design. The test also checks the confinement of the beam and that the edge of the image receptor is as close as possible to the chest wall edge, whereas the compression paddle is aligned with it.

This test may be also performed by using computed radiography plates larger than the image receptor area. In this case, by placing the computed radiography plate on the breast support and using adequate markers for the delineation of the light field, the coincidence of the light field with the radiation beam could be observed.

The 'missing tissue' on the chest wall side may also be determined by using an appropriate phantom having markers at its chest wall edge [18, 19].

The following equipment is required for the test:

(a) Computed radiography or film–screen system cassette bigger than 24 cm × 30 cm.
(b) Radiopaque tools (e.g. coins, wire).
(c) At least two radiographic rulers (with fine marks of ≤5 mm, and a separate marker at the midpoint, referred to as '0' mark). Alternatively, one ruler and another midpoint marker, such as a coin, could be used.
(d) Phosphorescent strips smaller than 20 mm × 50 mm.
(e) Metal foil or other similar radiopaque material to cover the phosphorescent strips.
(f) PMMA slabs of total thickness 45 mm.

4.2.2.3. *Procedure*

The test procedure is as follows:

(a) Tilt the gantry if needed to have a clear view on the surface of the breast support.
(b) Tape four phosphorescent strips overlapping the edges of the breast support as in Fig. 33.
(c) Place another piece of phosphorescent strip in the middle of the breast support. This is used as an indicator of X ray interaction.
(d) Place the metal foils on the strips to cover the parts that are on the breast support, except for one strip at the chest wall edge. Align that strip outside the perimeter of the chest wall by 5 mm — the tolerance for the missing tissue. The strip covered by this piece of foil will not be illuminated if the field alignment is correct.
(e) Darken the room to enable a clear view of the breast support, which is essential for this test.
(f) Set the largest field size and use the manual exposure mode to select exposure parameters that provide clear indications from the phosphorescent strips (e.g. 28 kV and 100 mA · s).
(g) Record whether any of the strips glow — indicating incorrect collimator setting. If necessary, repeat this exposure until none of the phosphorescent strips glows.
(h) Align and centre the PMMA slabs at the chest wall edge.
(i) Place a radiographic ruler perpendicular to the chest wall edge at its '0' mark to check the missing tissue at the chest wall edge.

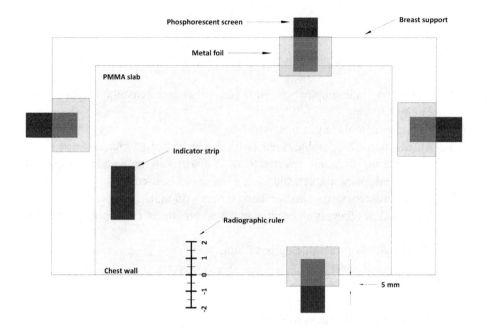

FIG. 33. Suggested layout for rulers and phosphorescent screen markers used to evaluate X ray field alignment and missing tissue.

(j) Place the radiographic ruler on the PMMA slabs at the patient contact edge of the compression paddle or use a coin to mark this position. A ruler provides a quantitative way to establish the amount of missing tissue at the chest wall edge.

(k) Apply a low compression force on the PMMA slabs.

(l) Make a manual exposure with settings that would be used for a standard breast 45 mm thick.

4.2.2.4. Analysis and interpretation

Examine the resulting image and determine the distance from the '0' mark on the ruler on the PMMA to the edge of the active detector. This is a measure of the missing tissue.

4.2.2.5. Baselines and tolerances

The following requirements apply:

(a) The missing tissue at the chest wall side should be <5 mm.

(b)   The X ray beam should completely irradiate the image receptor area but should not extend beyond the chest wall side by >5 mm.

(c)   The X ray beam on any other side of the image receptor area should not extend by >2% of the distance from the focal spot to the image receptor.

(d)   The chest wall edge of the compression paddle should not extend beyond the chest wall edge of the image receptor by >5 mm, and the chest wall edge of the paddle should not be visible in the image.

4.2.2.6.   Frequency

The test is repeated annually, and following service or replacement of the X ray tube, collimator or detector.

4.2.2.7.   Corrective action

If a misalignment is observed at the chest wall edge, then the use of the system should be suspended until repair, as this fault may impact clinical diagnosis. If the misalignment is present on other edges, then its extent should be evaluated and a decision should be made about whether the mammography unit may be used until the problem is resolved.

4.2.3.   Tube output

4.2.3.1.   Description and objective

The X ray tube output permits the calculation of the incident air kerma if tube loading (mAs value) and compressed breast thickness are known. Besides serving dosimetric purposes, the tube output is indicative of the radiation generator performance [4, 18, 20].

4.2.3.2.   Equipment

The following equipment is required for the test:

(a)   Dosimeter with calibrated ionization chamber or solid state detector.

(b)   Solid state detectors may also be used to evaluate the constancy of the equipment. Using a solid state detector opens the opportunity to determine other parameters with one exposure (e.g. exposure time, kilovoltage accuracy), if the instrument is calibrated for the radiation qualities used by the system.

(c)   Metal plate to shield the image receptor if it is not removable.

4.2.3.3. Procedure

The test procedure is as follows:

(a) Place the shielding plate on the image receptor if it is not removable.
(b) Place the instrument on the metal shield, ensuring that its sensitive area is centred laterally, 6 cm [18] or 4 cm [21] from the chest wall side.
Note: If a solid state detector with a rectangular sensitive area is used, then it should be parallel to the chest wall side to minimize the influence of the heel effect (see Fig. 34).
(c) Perform the measurement by removing the compression paddle to determine the incident air kerma.
Note: It is recommended that, at least at the time of commissioning, this measurement be repeated with the compression paddle in place to obtain information about its attenuation [20]. The compression paddle of some mammography systems is not removable. In this case, a forward scatter correction factor (1.076) could be used to determine the incident air kerma, or the compression paddle could be moved as far as possible from the instrument [20].
(d) Measure the distance between the focal spot and the sensitive area of the instrument.
(e) Choose a field size that is used clinically (e.g. 18 cm × 24 cm, 24 cm × 30 cm).
(f) Select the manual exposure mode and set the tube voltage according to Table 2. The current–time product should be high enough to have a reliable reading from the instrument (e.g. 20 mA · s).

FIG. 34. Set-up for measuring the tube output using a solid state detector.

TABLE 2. SUGGESTED KILOVOLTAGES FOR TUBE OUTPUT TEST

Selected kilovoltage (kV)	Anode–filter combination					
	Mo/Mo	Mo/Rh	Rh/Rh	W/Rh	W/Ag	W/Al
1	24	26	26	26	26	29
2	26	28	28	28	28	32
3	28	30	30	30	30	35
4	30	32	32	32	32	38

(g) Make more exposures by repeating step (f) at different tube voltages.
(h) Repeat steps (g) after changing to another radiation quality (anode–filter combination) until all clinically relevant combinations are tested.

4.2.3.4. *Analysis and interpretation*

Each instrument reading (e.g. temperature, pressure) should be corrected, if needed, and then multiplied by 1.076 to take into account the scattering effect of the compression paddle, if necessary [20]. Equation (1) describes the calculation of the tube output at the reference distance of 1 m.

The tube output is expected to decrease with time.

4.2.3.5. *Baselines and tolerances*

There is no limiting value for the tube output. The value 30 µGy/(mA · s) at a distance of 1 m was reported in earlier protocols as a minimum for Mo/Mo radiation quality.

4.2.3.6. *Frequency*

The test is repeated annually.

4.2.3.7. *Corrective actions*

If the tube output has changed significantly since the previous test or an unexpectedly large change has occurred, then check the kilovoltage accuracy, if possible, and measure the HVL value at 28 kV after repeating the test. If the system does not conform to the requirements, then contact a service engineer to discuss the possible issues. If the system fails the suggested tests, then it should not be used until it is repaired.

4.2.4. Half-value layer

4.2.4.1. Description and objective

The test determines the HVL, which is an indicator of the status of the X ray tube. Furthermore, the HVL needs to conform to certain limits according to national or international requirements [9, 18].

4.2.4.2. Equipment

The following equipment is required:

(a) Ionization chamber calibrated for mammographic radiation qualities;
(b) Ruler or measuring tape;
(c) High purity (>99.9%) aluminium filters;
(d) Metal plate to shield the image receptor if it is not removable.

4.2.4.3. Procedure

The test procedure is as follows:

(a) Place the metal shielding on the breast support if the image receptor is not removable. Otherwise, remove it from the bucky.
(b) Select the manual exposure and a target–filter–kilovoltage combination that would be selected by the AEC for 20 mm thick PMMA.
(c) Place the ionization chamber in a holder 45 mm above the breast support, centred laterally. Its central axis should be aligned at 40 mm [18] or 60 mm [21] from the chest wall edge.
(d) If possible, collimate the beam to be slightly larger than the sensitive area of the ionization chamber. If necessary, place a metal aperture in front of the collimator or on the compression paddle to achieve this.
(e) If the compression paddle is used, then placed it at half the focal spot–ionization chamber distance. Use of the paddle should be consistent with the tube output measurement.
(f) After making an exposure, record the measured values.
(g) Put a piece of aluminium slightly thinner than the expected one for the given target–filter combination on the compression paddle and repeat the

exposure (e.g. 0.3 mm of aluminium if the expected value according to the target–filter–kilovoltage combination is 0.35 mm). Check that the reading is more than half of that obtained without the filter. If it is not, use a thinner aluminium piece and repeat this procedure until a value is measured above the half value.

(h)   Add another piece of aluminium (e.g. 0.1 mm) and repeat the measurement. This reading should be less than half of the value measured without any attenuators. If it is not, increase the thickness of the attenuator until the ionization chamber measures less than half of the unattenuated reading.

(i)   As a precaution, repeat the measurement without any filters to check the unattenuated reading again.

(j)   Repeat steps (c)–(i) for every other target–filter–kilovoltage combination used clinically for standard (45 mm thick) and thick (70 mm thick) breasts.

Figure 35 shows a possible arrangement for the measurement.

4.2.4.4.   Analysis and interpretation

Calculate the value of the HVL using Eq. (2).

FIG. 35. Set-up for HVL measurement using an Al plate and an ionization chamber.

4.2.4.5. Baselines and tolerances

Acceptable HVL values are provided in Refs [9, 18] and can be calculated using Eq. (5):

$$\frac{V}{100}+0.03 \leq \text{HVL} \leq \frac{V}{100}+0.03+ C \tag{5}$$

where

C depends on the radiation quality and is equal to:
  0.12 for Mo/Mo;
  0.19 for Mo/Rh;
  0.22 for Rh/Rh;
  0.30 for W/Rh;
  0.32 for W/Ag;
  0.25 for W/Al;

and V is the nominal value of the tube voltage.

For other radiation qualities, refer to the manufacturer's specifications or the literature.

4.2.4.6. Frequency

The test is repeated annually.

4.2.4.7. Corrective actions

If the HVL is very low or very high, the tube voltage should be measured to confirm that the tube potential is properly calibrated. If the tube potential is appropriate, then a service engineer should be consulted. If the system fails the test and image quality is also deteriorated, then it should not be used until it is repaired.

4.2.5. Automatic exposure control system — reproducibility

4.2.5.1. Description and objective

The AEC system should deliver the radiation dose that is appropriate to achieve a given image quality for a defined test object. The image quality level and the parameter used to represent it are arbitrarily defined and depend on the phantom and the measurement method.

The AEC system should be consistent and provide constant settings for the same object; namely, it should be able to select the same technique factors for multiple exposures of the same test object and produce similar image quality [18, 20, 21].

4.2.5.2. Equipment

The following equipment is required for the test:

(a)   PMMA slabs with 45 mm total thickness;
(b)   High purity aluminium plate 0.2 cm thick and either 10 cm × 10 cm or 15 cm × 15 cm;
(c)   Ionization chamber for film–screen systems.

4.2.5.3. Procedure

The test procedure is as follows:

(a)   Computed radiography and digital radiography imaging systems:
  (i)   Place the PMMA slabs centred laterally on the breast support and align them to the chest wall edge.
  (ii)   Place the aluminium plate on top of the slabs, at 4 cm [18] or 6 cm [21] from the chest wall edge.
  (iii) Use a compression force that activates the AEC. Figure 36 shows the arrangement.
         Note: For the evaluation of these images, turn off every possible image processing device.
  (iv)   Select the AEC mode that is most frequently used clinically and, if possible, the AEC sensor that is aligned with the aluminium plate.
         Note: Always use the same computed radiography plate for this test and for each exposure to ensure that it is recently cleared.
  (v)   Make an exposure and note the feedback values from the operator's console.
  (vi)   Repeat step (v) four times to obtain a total of five images.
(b)   Film–screen systems:
  (i)   Place the PMMA plates on the breast support centred laterally and aligned with the chest wall edge.
  (ii)   Use a compression force that activates the AEC. Figure 36 shows the arrangement.

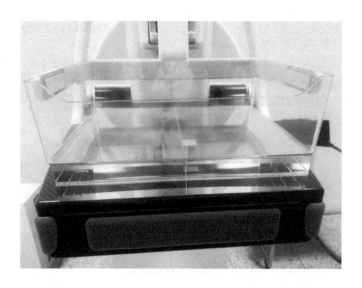

FIG. 36. Phantom set-up for the measurement of the signal difference to noise ratio using a 45 mm thick PMMA phantom and an aluminium plate.

  (iii) Select the AEC mode and, if possible, the AEC sensor that is 6 cm away from the chest wall edge.
     Note: The system may change between spectra with this AEC setting; in this case, it is advised to place another 5 mm thick PMMA plate on top of the slabs to effect the selection of a single spectrum.
  (iv) Place the ionization chamber as close as possible to the AEC sensor without causing interference to the AEC.
  (v) Make at least five exposures by repeating the previous steps. Record the feedback values from the operator's console and note the ionization chamber reading.

4.2.5.4. Analysis and interpretation

(a) Computed radiography and digital radiography imaging systems:
  (i) Use the image analysis software to calculate the MPVs and the standard deviation in each ROI, according to Fig. 37.
  (ii) Calculate the SDNR according to Eq. (6):

$$SDNR = \frac{\left|MPV_{bkg} - MPV_{Al}\right|}{SD_{bkg}} \tag{6}$$

where

MPV_{bkg} is the MPV of the ROI selected on the background;
MPV_{Al} is the MPV of the ROI selected on the aluminium plate;

and SD_{bkg} is the standard deviation of the pixel values of the background ROIs.

Short term reproducibility of the AEC can be assessed easily from the COV of the air kerma and the SDNR.

(b)   Film–screen systems:
Use the appropriate correction factors for the ionization chamber reading and then calculate the mean value of the results.

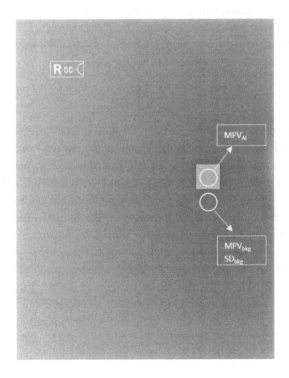

FIG. 37. Image of a test object and regions of interest used to calculate the signal difference to noise ratio.

4.2.5.5. Baselines and tolerances

The COV of the SDNR should be −5% to 5% in computed radiography and digital radiography systems.

The measured incident air kerma should differ by −5% to 5% from the mean for each exposure when film–screen systems are tested.

4.2.5.6. Frequency

The test is repeated every six months.

4.2.5.7. Corrective actions

If the SDNR is outside the tolerance range, then the cause should be investigated. Furthermore, consistency of the AEC system may be checked by performing the breast thickness compensation test. While the AEC system is not working properly, the system could be used in manual exposure mode if needed. Contact the service engineer to check the equipment.

4.2.6. Automatic exposure control system — breast thickness compensation

4.2.6.1. Description and objective

For mammography systems using digital image receptors to form the image, the SDNR is a more appropriate indicator than the dose used earlier for film–screen systems. Film–screen systems that are still in use, however, rely on the principle that a consistent dose should be delivered at the surface of the film–screen system, regardless of what objects are in the path of the beam. Because of this principle, the optical density of the image is the metric used for the evaluation of the AEC performance for film–screen systems.

This test serves the purpose of checking that the AEC system produces a consistent SDNR or that the optical density is constant on the film. It is advised that the test described in Section 4.2.1 be performed before proceeding with this one, in order to ensure that the thickness indication is correct [18, 20–22].

The following equipment is required:

(a) PMMA slabs with a total thickness of (at least) 20, 45 and 60 mm, covering the whole image receptor area;
(b) The aluminium plate used for the previous test (for digital mammography systems; Section 4.2.5.2);
(c) Densitometer (for film–screen systems).

4.2.6.3. *Procedure*

Several guidelines recommend testing of the thickness compensation for a range of thicknesses of the PMMA plates (e.g. 20–70 mm in 10 mm steps), including 45 mm. If the available capacities allow this, then it is advised to follow this recommendation.

The earlier IAEA guideline for QC in digital mammography [18] suggested the setting of equivalent breast thickness for PMMA (e.g. 53 mm indicated compressed breast thickness for 45 mm thick PMMA plates). This approach is not prohibited and may have a negligible impact on the results of the test; however, the choice of anode–filter combination relies on the indicated compressed breast thickness and it is required that QC tests relying on this be performed in a consistent manner.

The following test procedures are used:

(a) Computed radiography and digital radiography imaging systems:
   (i)   Place the 20 mm thick PMMA slab on the breast support, aligned with the chest wall edge.
   (ii) Place the aluminium plate on the PMMA slab at 4 cm [18] or 6 cm [21] from the chest wall edge, centred laterally. Figure 38 shows the set-up for different phantom thicknesses.
   (iii) After setting an appropriate compression force to operate the AEC system, select the most frequently used clinical AEC setting.
   (iv) If possible, select the AEC sensor that is closest to the aluminium plate.
   (v)   Expose the phantom in the AEC mode and note the feedback values from the operator's console. As a result of this exposure, the unprocessed image should be available for later evaluation.
   (vi) Repeat the previous steps without moving the aluminium plate; each time, increase the phantom thickness by placing the next slab under the previous one.
   (vii) Repeat the test for each AEC mode of operation used clinically.

FIG. 38. Phantom set-up for automatic exposure control measurement at different phantom thicknesses.

(b) Film–screen systems:
  (i) Place the 20 mm thick PMMA slab on the breast support, aligned with the chest wall edge.
  (ii) After setting an appropriate compression force to operate the AEC, select the AEC setting that is most often used clinically.
  (iii) Select the AEC sensor closest to the 6 cm spot from the chest wall edge. Select a clinically used density correction setting that would provide an optical density approximately equal to the target optical density (e.g. OD = 1.5, including fog and base).
  (iv) Expose the phantom in AEC mode and process the film.
  (v) Increase the phantom thickness and repeat the test.

4.2.6.4. Analysis and interpretation

The analysis and interpretation of the test results are carried out as follows:

(a) Computed radiography and digital radiography imaging systems:
  (i) Evaluate each phantom image and determine the SDNR for each, according to Section 4.2.5.
  (ii) Calculate the SDNR variation with respect to the reference thickness of 45 mm, $\Delta SDNR_{45}$, according to Eq. (7):

$$\Delta SDNR_{45} = \frac{SDNR_i - SDNR_{45}}{SDNR_{45}} \times 100 \tag{7}$$

where

$SDNR_i$ is the SDNR measured at the i-th thickness;

and $SDNR_{45}$ is the SDNR measured at 45 mm.

The AEC compensation performance may be evaluated in terms of $\Delta SDNR_{45}$.

(b) Film–screen systems:
Measure the optical density on each film using a densitometer.

4.2.6.5. *Baselines and tolerances*

The following baselines and tolerances apply:

(a) Computed radiography and digital radiography imaging systems:
The limiting values (for 20, 45 and 60 mm thick PMMA) are reported in Table 3. Reference [18] also provides achievable and acceptable SDNR values for different types of mammography units.
(b) Film–screen systems:
The optical density of all the films should not differ by more than OD = ±0.15 from the target optical density.

TABLE 3. BASELINES AND TOLERANCES FOR SDNR

PMMA thickness (mm)	Limiting values $\Delta SDNR_{45}$
20	$\geq 0\%$
45	—
60	$\geq -30\%$

4.2.6.6. Frequency

The test is repeated annually.

4.2.6.7. Corrective actions

First, check the exposure parameters and the geometry of the set-up, ensuring that the indicated breast thicknesses correspond to the expected tube voltage and anode–filter combinations.

In the case of digital mammography units, if the $\Delta SDNR_{45}$ is outside the limiting values, then check the image receptor. If the image receptor is performing properly, then the AEC may need adjustment. Corrective actions such as calibration should be taken before any further patients are imaged.

Film–screen systems may be affected by many factors. If the system fails only this test (e.g. processing is appropriate, tube output test is satisfactory), then a service engineer should be contacted.

4.2.7. Automatic exposure control system — consistency between sensors

4.2.7.1. Description and objective

The AEC devices in film–screen mammography units are usually located in several positions. This may be the case with some digital systems as well. When the AEC sensor (its position) can be selected, this test is necessary to ensure that the optical density or the target SDNR is consistent regardless of the chosen AEC sensor [18, 21].

4.2.7.2. Equipment

The following equipment is needed for the test:

(a) PMMA slabs with 45 mm total thickness;
(b) High purity aluminium plate with a thickness of 0.2 cm and a size of either 10 cm × 10 cm or 15 cm × 15 cm (for digital mammography systems);
(c) Densitometer (for film–screen systems).

4.2.7.3. Procedure

Follow the procedure described in Section 4.2.5. It is not necessary to repeat this test for each sensor position, but perform it at least once (at commissioning) for each of the selectable sensors. When repeating the procedure with a given sensor, ensure that the aluminium detail is at the position of the next selected sensor area.

It is recommended to use films instead of an ionization chamber for film–screen systems.

4.2.7.4. Analysis and interpretation

Analyse the images as described in Section 4.2.5.

4.2.7.5. Baselines and tolerances

For computed radiography and digital radiography systems, the SDNR should be less than $\pm 5\%$ of the mean. The measured optical density should not differ by more than $OD = \pm 0.20$ from the mean value measured at the AEC sensor area.

4.2.7.6. Frequency

The test is repeated every six months.

4.2.7.7. Corrective actions

When the result for any AEC sensor is outlying, then the AEC system should be recalibrated.

4.2.8. Operation of the automatic exposure control guard timer

4.2.8.1. Description and objective

Similarly to radiography, mammography units are equipped with a guard timer, which is activated if the AEC device is faulty and cannot terminate the exposure. The test described here should be performed at least at the time of commissioning to examine how the mammography unit behaves and what error message could be expected when the AEC is not working properly and the guard timer is triggered [18, 21].

4.2.8.2. Equipment

The following equipment is used in the test:

(a) A metal plate (1 mm steel or at least a 0.3 mm thick lead sheet) that can shield the AEC sensor;
(b) Optionally, an ionization chamber or a solid state detector to measure the incident air kerma.

4.2.8.3. Procedure

The test procedure is as follows:

(a) Place the metal plate on the breast support.
(b) Place the instrument on the metal plate, with its sensitive area centred laterally 4 cm [18] or 6 cm [21] from the chest wall edge.
(c) Use a compression force that activates the AEC. If necessary, use spacers (see Fig. 39).
(d) Select the AEC mode most often used clinically.
(e) Make an exposure and note the displayed values on the operator's console. If error messages appear after the exposure, also record them.

(a)   (b)

FIG. 39. Measurement set-up used to test the operation of the automatic exposure control guard timer, with spacers viewed (a) from above and (b) from the side.

4.2.8.4. Analysis and interpretation

Note how the system behaves. If it terminates after the pre-exposure and sends an error message, then it is acceptable, as the system prevents further radiation to the patient. Some systems perform the exposure and trigger the guard timer without further notice.

4.2.8.5. Baselines and tolerances

It is expected that the operation of the AEC does not change from the baseline (set during commissioning) over time.

4.2.8.6. Frequency

The test is performed at the time of commissioning.

4.2.8.7. Corrective actions

It is desirable that the system terminates the exposure if the AEC is faulty and that it displays a warning or, preferably, an error message. In some digital systems the control software can be set so that the AEC guard timer exhibits a different behaviour. Consult a service engineer about whether it is necessary and possible to change the mode of operation of the AEC guard timer.

4.2.9. Response function and noise evaluation (computed radiography and digital radiography only)

4.2.9.1. Description and objective

The purpose of this test is to evaluate the response and noise characteristics of the image acquisition system under standard conditions. Regardless of the type of imaging system, the MPV of the image as a function of dose should follow a trend, be it linear (digital radiography) or non-linear (computed radiography). An appropriately set system exhibits a behaviour where its quantum noise is the most dominant [18, 20].

4.2.9.2. Equipment

The equipment used for the test is PMMA slabs 45 mm thick.

FIG. 40. Phantom set-up for measuring the detector uniformity and noise.

4.2.9.3. Procedure

The test procedure is as follows:

(a) Place the 45 mm thick PMMA slab(s) on the breast support.
(b) Select a manual exposure mode on the operator's console.
(c) Compress the PMMA slabs and choose an anode–filter combination and a kilovoltage setting that would be set by the AEC for this set-up.
(d) Lower the compression paddle (see Fig. 40).
(e) Make an exposure with a fraction of the mAs value (e.g. 10 mA · s) that would be chosen by the AEC for this set-up. After the exposure, note the feedback values and the incident air kerma (or ESAK).
(f) Repeat the previous step with increasing mAs value (increase the value by a factor of approximately 1.4–1.6 for each step) to obtain a total of eight to ten images (reaching 200 to 300 mA · s).
Note: Always use the same computed radiography plate for the image acquisitions in this test and set the same reading mode while processing it.

4.2.9.4. Analysis and interpretation

The following processes are used for the analysis and interpretation of the test results:

110

(a) Response function:

    (i)   Use image analysis software to evaluate the images by measuring the MPV and standard deviation in a ROI (approximately 1 cm^2) placed in the centre, 6 cm from the chest wall edge. For computed radiography systems, also note the exposure index.

        Note: Use the same ROI parameters for each image.

    (ii)   Plot the MPV versus the incident air kerma for digital radiography systems. For computed radiography systems, refer to the manufacturer's user manual about whether a linear or logarithmic relationship applies between the MPV and the incident air kerma or mAs value, and correct the measured or feedback values accordingly (e.g. take the natural logarithm of the incident air kerma if such a relationship is valid).

    (iii) Determine the best fit of the plot using least squares regression analysis and calculate the R^2 parameter. The line parameters follow Eq. (8):

$$\text{MPV} = A + Bx \qquad\qquad\qquad (8)$$

where

A is the incident point of the line on the y axis (pixel offset value);

B is the slope of the line;

and x corresponds to the incident air kerma or its logarithm.

    (iv) The mean distance of the measured results (the COV) from the fitted line can also be evaluated, by using Eq. (9):

$$\text{COV} = \frac{\text{MPV}_i - A}{x_i} \qquad\qquad\qquad (9)$$

where

MPV_i is the MPV of the i-th image;

and x_i corresponds to the incident air kerma or its logarithm for the i-th image.

(b) Noise evaluation:

The noise of an image can be characterized by the standard deviation of the pixel values of an image. The variance (standard deviation squared) is plotted against the mAs value or the incident air kerma.

The pixel values require normalization for computed radiography systems. The reciprocal of the mAs value or the incident air kerma can be used for linear fitting and to determine the correlation coefficient, R^2, which can be used to evaluate the noise through Eq. (10):

$$SD^2 = a + bx \tag{10}$$

where

    SD   is the standard deviation;
    a is the offset parameter;
    b is the slope of the line;

and x corresponds to the mAs value, the incident air kerma (for digital radiography) or the reciprocal of these parameters (for computed radiography).

4.2.9.5. Baselines and tolerances

The correlation coefficient (R^2) should be as close as possible to 1 — specifically, R^2 should be ≥ 0.95 [20] or ≥ 0.99 [21]. The COV for the response function should be <10%.

4.2.9.6. Frequency

The test is repeated annually.

4.2.9.7. Corrective actions

It is advisable to review the results and repeat the test, or at least repeat the evaluation. If the image response function shows deviations, then ensure that the radiation output and the HVL are satisfactory. Consult a service engineer if the previously mentioned QC tests (Section 4.2.9) are not acceptable.

Increasing noise of an imaging system may be due to several issues. If the system in question uses a digital radiography image receptor, then it may be adjusted, calibrated or eventually exchanged. If increased noise is observed in a computed radiography system, then the reader units and the imaging plates should be checked thoroughly. If further tests confirm that the deterioration of the response function impacts image quality, then suspending the system should be considered.

4.2.10. Image receptor uniformity — annual test (computed radiography and digital radiography only)

4.2.10.1. Description and objective

The test assesses the 'flatness' of the image. A homogeneous test object exposed on a digital image receptor should produce a uniform image. This test is essential for digital mammography systems, as it determines whether a flat field correction is necessary or the gain correction of individual pixels is correct (digital radiography only) [20, 21].

4.2.10.2. Equipment

A 45 mm thick PMMA slab covering the whole image receptor area is needed for the test.
Note: Ensure that the PMMA slab is uniform and has no deep scratches or inhomogeneities.

4.2.10.3. Procedure

The test procedure is as follows:

(a) Place the PMMA slab on the breast support.
(b) Compress the test object, choose the largest compression paddle available to effect the selection of the largest available field.
(c) Select the most frequently used AEC mode to make an exposure and note the feedback values from the operator's console.
   Note: Use the same computed radiography plate that was used for the response function evaluation. This test should be repeated for each computed radiography plate if sensitivity matching is performed.

If the standard slab has suspected imperfections or visible scratches, the slab may be rotated and the test repeated to avoid artefacts arising from this.

4.2.10.4. Analysis and interpretation

It is advised to use image analysis software for this test, such as the 'COQ Mammo' plug-in for ImageJ developed by the European Federation of Organisations for Medical Physics (EFOMP) [20]. The following steps are followed to determine the local and global uniformity of an image. Global uniformity is only applicable for digital radiography systems in which the flat field correction is performed for the images.

(a) Use image analysis software to determine the MPV on the whole image.
(b) Place ROIs 10 mm^2 or larger around the image and next to each other, forming a matrix (see Fig. 41).
(c) Determine the MPV for each ROI, as well as for its eight neighbouring ROIs.
(d) Calculate the local uniformity (LU) according to Eq. (11), which determines the ROI with the largest relative difference from its neighbouring ROIs ('max') as a metric of local uniformity:

$$LU = \frac{\left| MPV_{i,j} - MPV_n \right|}{MPV_n} \tag{11}$$

where

$MPV_{i,j}$ is the MPV of a ROI of the image, where i and j denote the position of the ROI on the image;

and MPV_n is the MPV of the neighbouring ROIs, indexed with n.

Note: Inherent non-uniformities, such as regions or edges where the image is inherently inhomogeneous, should be omitted.

(e) Global uniformity (GU) is a similar measure of uniformity, but the MPV of the whole image serves as a basis for comparison with the baseline values or with $MPV_{i,j}$, as described by Eq. (12):

$$GU = \frac{\left| MPV_{i,j} - MPV_{image} \right|}{MPV_{image}} \tag{12}$$

where

MPV_{image} is the MPV of the whole image.

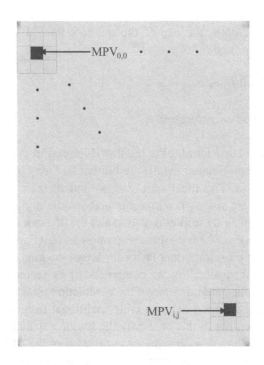

FIG. 41. Positions of ROIs used for the assessment of local and global uniformity.

4.2.10.5. Baselines and tolerances

Local uniformity should be ≤5% and global uniformity (digital only) should be ≤10%.

4.2.10.6. Frequency

The test is repeated annually or when changes are made to the imaging system.

4.2.10.7. Corrective actions

Repeat the test and check the PMMA slabs for inhomogeneities. Rotate the slabs and repeat the test. Consult a service engineer if dead pixels are visible or a recalibration is necessary. Consult the radiologists using this system to ensure that an observed error does not significantly impact clinical evaluation of the images; if it does, the use of the machine should be suspended until recalibration is performed.

4.2.11. Spatial resolution

4.2.11.1. Description and objective

This test is used to determine the smallest (high contrast) detail that is visible by an imaging system. Smaller details are blurred and may not be distinguished from the background. The focal spot size — and in the case of film–screen systems, the film grain size — is a limiting factor, while digital imaging systems are affected by the imaging technology used and the detector elements.

Films have superior high contrast resolution to digital imaging devices, but this is not the single most important factor for breast imaging and diagnosis.

With the wide availability of computers, it is recommended to use a special test tool for the determination of the modulation transfer function (MTF). This test may also be carried out in a more traditional fashion, with a line pair resolution test object, but in this case only the limiting spatial resolution can be determined. This section describes the determination of the MTF. The test using a line pair resolution tool differs only in the test object itself and its subjective evaluation [18, 21].

4.2.11.2. Equipment

The following equipment is needed:

(a) An appropriate tool to obtain the MTF. This test tool usually consists of a very precisely manufactured metal attenuator with a very straight edge, about 0.1 mm thick, and constructed from materials that are easily manufactured, durable and have a high density and attenuation, such as niobium, steel or tungsten. High contrast is necessary for the image of this object; therefore, high attenuation is required. Some manufacturers of such test objects use slightly angled lead or a thicker aluminium plate, both of which are more malleable; in this case, the MTF test tool is usually an insert in a PMMA baseplate.
(b) For film–screen systems, a line pair resolution tool.
(c) PMMA slabs 45 mm thick.
(d) MTF evaluation software.

4.2.11.3. Procedure

The test procedure is as follows:

(a) Place the MTF tool 45 mm above the breast support so that the metal attenuator is angled slightly (2°–5°) with respect to the chest wall edge Figure 42 shows the arrangement. In the case of film–screen systems, use a phantom with a bar pattern to determine the spatial resolution.
(b) Make an exposure setting manually exposure factors similar to those used to obtain a clinical image of the average breast. Figure 43 shows an image obtained by such an exposure.

4.2.11.4. Analysis and interpretation

The following process is used for the analysis and interpretation of the test results:

(a) Process the image using the appropriate software, which calculates the line spread function from the transition of the slanted edge (first derivative), performs a fast Fourier transform and processes the function to plot it against the spatial frequency.

FIG. 42. Set-up for the determination of the modulation transfer function using a 10 mm thick phantom placed above a 35 mm PMMA slab.

FIG. 43. X ray image of the modulation transfer function test tool (right) mounted in a quality control test object.

(b) Note the points where the MTF falls below 50% and 20%.
(c) During commissioning, establish a baseline and record the transfer rate at 2.5, 5 and 7.5 cycles per millimetre.

4.2.11.5. Baselines and tolerances

The following should be considered:

(a) The spatial frequencies at which the MTF drops to 50% and 20% should not be less than the values specified for the relevant model.
(b) The MTF at 2.5, 5 and 7.5 cycles per millimetre should not differ from the baseline by more than 10%.
(c) The spatial resolution may vary on different regions of the image receptor if it deteriorates. In this case, the measurement should be repeated at several locations on the image. These images shall then be compared to each other.
(d) Higher noise may cause the MTF to vary locally.
(e) When a line pair resolution tool is used to determine the limiting spatial resolution, refer to the manufacturer's specifications.
(f) For film–screen imaging, the limiting spatial resolution should be better (higher) than 12 line pairs per millimetre.

4.2.11.6. Frequency

The test is repeated annually, and after changes or service to the detector, tube or computed radiography plate reader.

4.2.11.7. Corrective actions

If the 50% or 20% frequencies for the MTF, either parallel or perpendicular to the chest wall, drop below the manufacturer's specifications or if the MTF at 2.5, 5 and 7.5 cycles per millimetre has changed by 10% or more from previously measured values, consult the manufacturer's service representative.

When the limiting spatial resolution is less than the desirable value, confirm that the system is performing satisfactorily and that processing is acceptable for film–screen systems, and check the radiation output. The focal spot size may affect the resolution.

4.2.12. Ghosting (computed radiography and digital radiography only)

4.2.12.1. Description and objective

This test evaluates the severity of artefacts caused by previous exposure to the detector in all systems except photon counting systems [18].

4.2.12.2. Equipment

The equipment required is 45 mm thick PMMA slabs.

4.2.12.3. Procedure

(a) Place the PMMA slab on the right half of the breast support so that approximately one half of the breast imaging area is covered. The set-up is shown in Fig. 44.
(b) Compress the PMMA slab and then make a manual exposure with the parameters used for the average breast. Annotate the image as 'ghost 1' or similar.
   Note: When performing this test on computed radiography systems, immediately process the computed radiography plate and reuse the same plate for this test.
(c) Reposition the slab on the breast support so it is centred laterally and aligned to the chest wall edge (Fig. 45).

(d) After the first image is acquired, as soon as the unit allows another exposure or the same detector plate is available, acquire a second image with the same manual technique. This is annotated as 'ghost 2'.

Note: For computed radiography systems, let the same time elapse between making the first image and the following one.

FIG. 44. Phantom set-ups for the creation of 'ghost 1' and 'ghost 2' images.

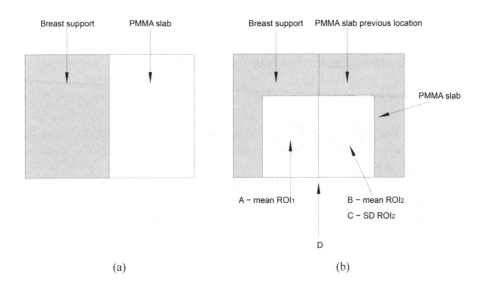

FIG. 45. Schematic representation of the ghosting test set-up for (a) the first and (b) the second exposure. ROI: region of interest; SD: standard deviation.

4.2.12.4. Analysis and interpretation

The process to analyse and interpret the test results is as follows:

(a) View the acquired images on the image display used for diagnosis. Draw two ROIs (area ~4 cm^2): one positioned on the part of the image that was covered by the PMMA slab only during the second exposure (ROI$_1$) and one on the part of the image that was covered by the PMMA slab during both exposures (ROI$_2$). Measure the MPV in ROI$_1$ (*A*), the MPV in ROI$_2$ (*B*) and their standard deviation (*C*).
(b) Record the results.
(c) Calculate the SDNR of the ghost image using Eq. (13):

$$\text{SDNR} = \left| \frac{A - B}{C} \right| \tag{13}$$

where

A is the MPV in the background of the phantom on the side where no attenuator was present during the first exposure. The ROI should be ~20 mm to the left of the line indicating the centre of the image receptor, denoted with 'D' in Fig. 45.

B is the MPV in the phantom's image, which is the background. This is the place where the PMMA slab was present during the first exposure. The ROI should be ~20 mm from line 'D'.

C is the standard deviation for the same ROI as for *B*.

(d) Use a narrow window width and appropriate window level, inspect the central part of the image where the boundary between the two areas of exposure lies. Record the window width and window level.
(e) Record the presence or absence of any visually observable ghost image.

4.2.12.5. Baselines and tolerances

An SDNR $\leq$ 2.0 for the ghost image is considered. Alternatively, when viewing the measured ghost image with a window setting typical of clinical use, there should be no visible indication of the first location of the attenuator in the second image.

4.2.12.6. Frequency

The test is repeated annually and after replacing the detector.

4.2.12.7. Corrective actions

When ghosting is present, evaluate the computed radiography plates to determine whether they are still acceptable in clinical practice. With time, these imaging devices deteriorate. When ghosting occurs in digital radiography systems, contact the supplier or the service representative to inspect the image receptor. If persistent ghosting is present on any type of image receptor, it should be suspended from clinical use.

4.2.13. Computed radiography plate sensitivity matching

4.2.13.1. Description and objective

Since each computed radiography plate is an individual image receptor, the tests investigating the imaging system should be performed for each one. As such tests are difficult to carry out in practice, it is considered acceptable to perform the QC tests with a single 'reference' computed radiography plate. Computed radiography plates from different series should not be mixed, as they may have significant differences in their response.

Before performing this test, ensure that the AEC is operating properly [18, 20].

4.2.13.2. Equipment

PMMA slabs of 45 mm thickness are needed for the test.

4.2.13.3. Procedure

The following procedure is used:

(a) Place the PMMA slab on the breast support, as in the uniformity test described in Section 4.2.10.
    Note: If the standard slab has suspected imperfections or visible scratches, the slab may be rotated and the test repeated to avoid the generation of artefacts.
(b) Apply compression to the test object, using the largest compression paddle available to effect the selection of the largest available field.

(c)   Select the most frequently used AEC mode, make an exposure
        and note the feedback values from the operator's console. If it is
        possible, select an AEC sensor close to a reference point 4 cm [18]
        or 6 cm [21] from the chest wall edge, centred laterally.
        Note: Do not change the AEC mode or the sensor throughout this test.
(d)   Make an exposure with the selected settings and record the feedback
        values from the operator's console (incident air kerma) and the computed
        radiography reader (exposure index).
(e)   Repeat the test for each computed radiography plate.

4.2.13.4. Analysis and interpretation

The analysis and interpretation of the test results are carried out as follows:

(a)   Visually evaluate the images on an image display used for diagnosis. Record
        whether any artefacts are present.
(b)   Using image analysis software, draw an ROI (~1 cm^2) at the reference
        point — 4 cm [18] or 6 cm [21] from the chest wall edge — in the
        centre of the first image and measure the mean and the standard
        deviation of the pixel values in the ROI. Record the results.
        Note: Since this test consists of imaging the PMMA slabs and it is carried
        out in a similar manner as the homogeneity test, it is advisable to perform
        the evaluation described in Section 4.2.10 for each computed radiography
        plate as well.
(c)   Repeat step (b) for each image.
(d)   Calculate the signal to noise ratio (SNR) for each image i using Eq. (14):

$$SNR_i = \frac{MPV_i}{SD_i} \tag{14}$$

(e)   For each parameter (SNR, MPV and exposure index), calculate the largest
        difference for each image from the reference using Eq. (15):

$$\Delta_{ref} = \max\left(\frac{|X_i - X_{ref}|}{X_{ref}}\right) \times 100 \tag{15}$$

where

Δ_{ref} is the variability with respect to the reference computed
                   radiography plate;
X_i is the exposure index, MPV or SNR of the i-th image;

123

and X_{ref} is the exposure index, MPV or SNR, correspondingly, of the image recorded with the reference computed radiography plate.

(f) For each parameter (SNR, MPV and exposure index), calculate for each image the largest difference from the mean value of all images using Eq. (16):

$$\Delta_{mean} = \max\left(\frac{|X_i - X_{mean}|}{X_{mean}}\right) \times 100 \tag{16}$$

where

Δ_{mean} is the variability with respect to the mean value of all computed radiography plates;

and X_{mean} is the mean value of the exposure index, MPV or SNR, calculated from all images.

4.2.13.5. Baselines and tolerances

The maximum deviation should be within ±15% of the mean of all images and within ±20% of the MPV, SNR and exposure index of the reference image. There should be no artefacts present on any image that would impact clinical diagnosis.

Reference [18] contains further advice on tolerances for specific makes of mammography systems.

4.2.13.6. Frequency

The test is repeated annually or when new cassettes are commissioned.

4.2.13.7. Corrective actions

Plates that have defects or do not perform as expected should be removed from use, as they may impact diagnosis. Artefacts present on images that cannot be removed indicate serious deterioration of the imaging plate. Such plates should be replaced. Every computed radiography plate has a limited useful life cycle, which depends on how extensively the plate is used, and should be replaced when not working as expected.

4.2.14. Mean glandular dose

4.2.14.1. Description and objective

In mammography, the radiation sensitive breast tissue that is exposed is glandular. This test determines the mean glandular dose (MGD), which is also displayed on the operator's console or may be available in the image's meta-information in digital systems. Determination of the incident air kerma, described in Section 4.2.3, is essential to calculate the MGD [18, 20, 21].

4.2.14.2. Equipment

The following equipment is needed for the test:

(a) PMMA slabs of thickness 20, 30, 40, 45, 50, 60 and 70 mm and, optionally, 80 mm;
(b) Spacers to set the indicated breast thickness to be equal to the equivalent breast thickness;
(c) Dosimeter with calibrated ionization chamber or solid state detector for the anode–filter combinations used by the mammography system.

4.2.14.3. Procedure

Two methods can be used to determine the MGD. The first method uses PMMA plates that are placed on the breast support and determines the AEC parameters and the measurements simultaneously. Using this approach, a PMMA slab with a given thickness (e.g. 20 mm) is placed on the breast support, compression for the equivalent breast thickness (21 mm for this example) is applied, and an exposure is made while the ionization chamber is placed above the compression paddle, where it does not interfere with the AEC sensor (Fig. 46). The instrument reading is recorded, corrected for backscatter and distance if necessary, and then the incident air kerma and MGD are calculated.

The second method, which is universal and usable for every type of imaging system, is called the 'substitution' method. This method first determines the AEC settings and then measures the tube output, as follows:

(a) Place the 20 mm thick PMMA slab on the breast support.
(b) Use the compression paddle to set a force that activates the AEC. Ensure that the indicated breast thickness corresponds to the equivalent breast thickness.
Note: When a setting is made at the borderline where the AEC would change

to a different anode–filter combination (e.g. 50 mm PMMA where 10 mm spacing is required, resulting in a total of 60 mm), ensure that the displayed breast thickness corresponds to a value for which the same anode–filter combination would be chosen. In this example, the system may choose a different anode–filter combination at 61 mm, but not at 59 mm thickness.

(c) Make an exposure using the most frequently used AEC settings (sensor area and mode).

(d) Record the feedback values of the exposure, including the anode–filter combination, kilovoltage and mAs value, as well as the ESAK and MGD calculated by the system, if available.

(e) Repeat the previous steps for the other thicknesses of PMMA, setting the appropriate equivalent breast thickness for each one.

(f) Refer to Section 4.2.3 for guidance on the tube output measurement. Set the parameters that were recorded in step (d). As an example, if the AEC would set 28 kV, W/Rh and 127 mA · s for a 45 mm thick PMMA plate (53 mm thick standard breast), then perform the tube output measurement with manual exposure settings of 28 kV, W/Rh and 20 mA · s and then correct the results. Subsequently, this reading should be corrected and used for the calculation of the MGD.

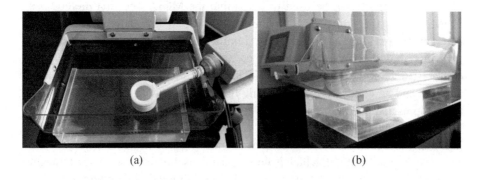

(a)   (b)

FIG. 46. Set-ups for the different methods used to determine the MGD and the AEC parameters. (a) Ionization chamber placed above the compression paddle; and (b) PMMA plates placed on the breast support.

4.2.14.4. Analysis and interpretation

The analysis and interpretation of the test results are carried out as follows:

(a) Calculate the incident air kerma from the corrected reading of the instrument at the given mAs value for the entrance surface of the breast.
   Note: As an example, for a focal spot–breast support distance of 700 mm and a 20 mm PMMA slab, the reference distance is 679 mm, as it is the difference between the focal spot–breast support distance and the equivalent breast thickness corresponding to 20 mm thick PMMA (Table 4).

(b) Calculate the MGD for each phantom thickness according to Eq. (17):

$$\text{MGD} = K_{i,t} \times g_t \times c_t \times s_t \tag{17}$$

where

$K_{i,t}$ is the incident air kerma without backscatter at the entrance surface of an equivalent breast thickness of t mm;

g_t is the factor used to convert air kerma to MGD for a breast having a 50:50 composition of glandular to fatty tissue and a thickness of t mm;

c_t is the conversion factor used to correct for the glandularity difference from the standard breast at a thickness of t mm;

and s_t is the factor used to correct for the radiation quality.

4.2.14.5. Baselines and tolerances

Table 4 lists the acceptable and achievable levels of MGD [18]. The acceptable level should be considered as a limit at which corrective actions should take place, while achievable levels are quoted here only for information.

4.2.14.6. Frequency

The test is repeated annually.

4.2.14.7. Corrective actions

Observe the variation of MGD over time. If the acceptable levels are exceeded, investigate the operation of the radiation generator and the AEC. If necessary, consult with a service engineer on the root cause of the issue. Computed

TABLE 4. ACCEPTABLE AND ACHIEVABLE MGD LEVELS [18]

Thickness of PMMA (mm)	Equivalent breast thickness (mm)	Acceptable MGD level (mGy)	Achievable MGD level (mGy)
20	21	<1.0	<0.6
30	32	<1.5	<1.0
40	45	<2.0	<1.6
45	53	<2.5	<2.0
50	60	<3.0	<2.4
60	75	<4.5	<3.6
70	90	<6.5	<5.1
80	103		

radiography plates used extensively may age and could require a higher radiation dose to provide the same image quality as earlier. If the generator has been adjusted to cope with ageing plates, then it will result in increased MGD.

4.2.15. Subjective evaluation of image quality

4.2.15.1. Description and objective

Subjective assessment of image quality is recommended in Section 4.1.2. In order to set a baseline, it is recommended that the test be repeated several times [18, 20, 22].

4.2.15.2. Equipment

The following equipment is needed for the test:

(a) Nationally or internationally recommended phantom with low contrast objects used by the facility for its QC programme for film–screen or digital mammography systems;

(b)   Image quality test objects (e.g. filaments, microcalcifications, discs);

(c)   Insets for the image quality test object;

(d)   The baseline phantom image from a previous QC test.

4.2.15.3. Procedure

The test procedure is detailed in Section 4.1.2.

4.2.15.4. Analysis and interpretation

The phantom image should be evaluated as described in Section 4.1.2.

4.2.15.5. Baselines and tolerances

Since the purpose of this test is to set the baseline for image quality indicators, it is advisable that the test be repeated several times to obtain several (e.g. five) images from which a more rigorously set baseline can be determined. Depending on the number and type of inset for the image quality test object (e.g. filaments, microcalcifications, discs), a scoring scheme should be defined.

4.2.15.6. Frequency

The test is repeated annually.

4.2.15.7. Corrective actions

If the image quality deteriorates over time, it is necessary to carry out further investigation (e.g. kilovoltage, AEC, display and processing algorithms) to determine the source of the change. If the problem persists, contact the service engineer.

5. COMPUTED TOMOGRAPHY

QC tests for CT are mainly intended to verify the operational stability of the equipment, after acceptance and commissioning tests have been performed. According to their priority, tests are classified into two types: essential and desirable [23]. Tests that may have a direct impact on clinical diagnosis and have an important role in revealing errors in imaging are considered essential.

Other CT performance tests, such as for kilovoltage accuracy and HVL, are not considered necessary for basic QC purposes.

5.1. QUALITY CONTROL TESTS FOR RADIOGRAPHERS

5.1.1. Daily startup procedure

CT manufacturers usually require a daily startup procedure to calibrate the detectors and check the operability of the system. This procedure is usually automated, and specific manufacturer's instructions should be followed. It also includes tube warmup and other procedures according to the given manufacturer's requirements [23].

5.1.2. Computed tomography laser alignment beams

5.1.2.1. Description and objective

Every CT system is equipped with laser lights to guide the alignment of the patient, one of the first steps before imaging. The laser alignment system is not restricted to the CT system itself, as CT systems used for radiotherapy often have ancillary laser systems for positioning. This test aims to confirm that the internal and external laser alignment beams are properly aligned with the tomographic plane [23].

5.1.2.2. Equipment

The test device consists of a thin absorber (e.g. a straight wire with a diameter of about 1 mm) positioned on a standard CT phantom, or another object that will give a high contrast when imaged. The test device comprises a PMMA base plate with two pegs connected with the straight wire. The pegs have vertical and horizontal holes of 1 mm and are spaced 25 cm from each other, arranged so that the wire is parallel to the tomographic plane, as shown in Fig. 47.

5.1.2.3. Procedure

The test procedure is as follows:

(a) Align the laser alignment test device centrally with the outer (or external) laser beam. Ensure that the test object (the wire) is parallel to the tomographic plane.

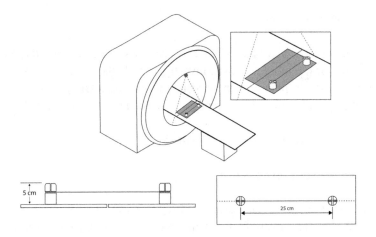

FIG. 47. Laser alignment test device.

(b) Check the test device's alignment with the internal laser beams.
(c) Perform a scan with narrow axial slices around the test object (±3 slices). Use a slice width corresponding to the diameter of the wire (e.g. 1 mm).

5.1.2.4. Analysis and interpretation

Determine the deviation between the expected vertical position and the location of the thin absorber by comparing the images of the series. If the wire and the lasers are aligned properly, then the entire wire should be in the same imaging plane.

5.1.2.5. Baselines and tolerances

Follow the manufacturer's specifications; otherwise, a deviation of ±5 mm is acceptable. For radiotherapy applications, it is recommended to set a lower tolerance.

5.1.2.6. Frequency

The test is repeated monthly.

5.1.2.7. Corrective actions

If the internal or external light fields are misaligned, contact the medical physicist to perform a more thorough investigation. If the CT system is used for radiotherapy treatment planning, then misalignment and a larger deviation could mean that the operation must be suspended until the lasers are realigned.

5.1.3. Scan projection radiograph accuracy

5.1.3.1. Description and objective

The scan projection radiograph (SPR) is used by the radiographer to prescribe the start and finish of a CT acquisition series, as well as for measuring distances. (Manufacturers' terminology for the SPR includes scoutview, scanogram, topogram, surview and pilot) The objective of this test is to ensure that the SPR image indicates the patient position accurately [23].

5.1.3.2. Equipment

A phantom or material block of a precisely known length (at least 25 cm, and preferably 50 cm or longer; see Fig. 48) is required for the test. The SPR distance can be determined by comparing the distances indicated between the ends of the test object.

5.1.3.3. Procedure

The test procedure is as follows:

(a) Obtain an axial acquisition with 1 mm (or the thinnest available) slice width at each end of the scan sequence or 1 mm (or the thinnest available) reconstructed display width.
(b) Place the SPR test object on the couch longitudinally.
(c) Make an SPR image, ensuring that the markers indicating the edges of the phantom are included.
(d) Make axial scans around the indicators at each end of the phantom (e.g. ±3 slices).

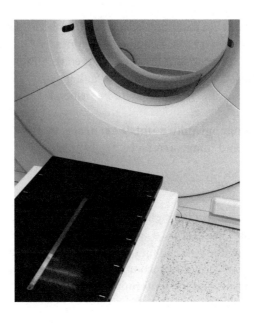

FIG. 48. A 50 cm long steel ruler set to measure the accuracy of the scan projection radiograph.

5.1.3.4. Analysis and interpretation

The following steps are taken to analyse and interpret the results:

(a) Check that each CT slice acquired around the markers based on the SPR image is centred over each marker.
(b) Determine the imaged distance between each pair of markers and compare it with the expected value.

5.1.3.5. Baselines and tolerances

Deviations of ±2 mm from the expected value are acceptable.

5.1.3.6. Frequency

The test is repeated every six months.

5.1.3.7. Corrective actions

If the measurements exceed the suggested tolerance, then repeat the test and ensure the correct geometrical positioning of the test object. If the test fails

133

again, then contact the medical physicist. If the problem persists, the medical physicist needs to contact a service engineer for corrective actions. If a consistent deviation from the expected value is observed, then recalibrate the display. If the deviation is not consistent, then a mechanical fault may be present. Depending on the severity of the issue, the machine could be suspended from service.

5.1.4. Computed tomography number accuracy, image noise, image uniformity and image artefacts

5.1.4.1. Description and objective

It is necessary to regularly check the accurate representation of a water equivalent test object, as it is used as a basis for image formation. To ensure that the images are uniform, scanning of a water filled test object (or a phantom containing uniform material) is recommended. This test investigates the CT number and noise levels and the uniformity of CT images to verify that they are within tolerances and that no image artefacts are visible [23, 24].

5.1.4.2. Equipment

The manufacturer's phantom, a commercial phantom or a simple round phantom made of water equivalent material is required for this test.

5.1.4.3. Procedure

A CT protocol specified by the physicist at commissioning or the manufacturer's QC protocol should be used; otherwise, the manufacturer's specifications may be used. The test is performed by centring the phantom in the tomographic plane and scanning it using the specified protocol.

5.1.4.4. Analysis and interpretation

The analysis and interpretation of the test results are carried out as follows:

(a) Measure the CT number accuracy, image noise and uniformity, and confirm the absence of image artefacts on the same phantom images.
(b) Measure the CT number and image noise in a centrally placed circular ROI of appropriate diameter, as specified at commissioning. The CT number is the measured ROI mean value, and the noise is the standard deviation within the ROI (Fig. 49).

(c)   Draw five circular ROIs: four near the perimeter (about 1 cm from the perimeter) and one in the centre. The ROI diameters should be about 10% of the phantom's diameter and their spacing should correspond to the 12, 3, 6 and 9 o'clock positions on a clock's face.

(d)   Determine the uniformity as the absolute difference between the CT numbers in the centrally placed ROI and those in each of the four equivalent ROIs near the perimeter. Compare each of these four values with the given tolerances.

(e)   To measure the noise, draw one larger ROI centred in the phantom with a diameter of about 40% of the phantom.

(f)   Visually inspect all images acquired during the noise test for image artefacts.

5.1.4.5.   Baselines and tolerances

The following baselines and tolerances apply:

(a)   For CT number accuracy: ±5 HU (Hounsefield units; for water) from the baseline value is acceptable, or within the tolerance given by the manufacturer (if available). For other materials, refer to the phantom specifications and the baseline.

(b)   For image noise: ±25% of the baseline value is acceptable. Otherwise refer to the tolerance provided by the manufacturer (if available).

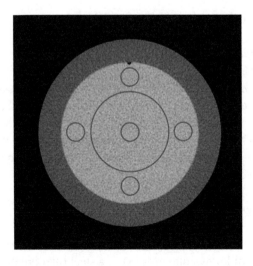

FIG. 49. Phantom image showing the positions of the ROIs used for the CT number accuracy, noise and uniformity tests.

(c)   For uniformity: ±10 HU (for water) is acceptable. If the manufacturer has provided a tolerance, then this should be used.

(d)   No artefacts that have the potential to compromise diagnostic confidence should be visible on any image.

The tolerances for CT number and image noise may be specified by the physicist at the time of commissioning.

5.1.4.6.   Frequency

For uniformity, image noise and image artefacts, the tests should be repeated weekly. The test for CT number accuracy should be repeated every six months.

5.1.4.7.   Corrective actions

The measured values may be outside of tolerances owing to several faults. Artefacts may arise from faulty image detector elements. Low exposure parameters may cause higher noise, while malfunction of the hardware or software components may also result in non-conformance. In such cases, user calibration may help; if this fails, the service engineer should be contacted. If a ring shaped artefact is present on one or several consecutive images, a defective detector element or detector row may be the cause. If recalibration does not resolve the inhomogeneity issues, then system use should be suspended.

5.1.5.   Accuracy of measured dimensions

5.1.5.1.   Description and objective

The aim of this test is to ensure that the distances measured on the image correspond to the actual distances. This test also evaluates the accuracy of the electronic distance indicator. While in most cases geometrical distortion does not have an impact clinically, accurate representation of the position of organs is indispensable when the system is also used for treatment planning and interventional procedures [4, 12].

5.1.5.2.   Equipment

Attenuating objects of known dimensions (i.e. coins attached to the phantom or inserts of known dimensions) or a steel ruler are needed for this test.

5.1.5.3. Procedure

Centre the phantom in the tomographic plane and scan it using the protocol determined by the medical physicist.

5.1.5.4. Analysis and interpretation

Measure the dimensions of the test objects using a measurement tool on the workstation and compare them to real dimensions and distances (see Fig. 50).

5.1.4.5. Baselines and tolerances

The measured dimensions should be within ±2%, and preferably within ±1%, of the nominal values.

5.1.4.6. Frequency

The test is repeated every three months.

5.1.4.7. Corrective actions

If the measured results are outside the tolerance range, then repeat the test by completely resetting the set-up. If deviations consistently appear on the

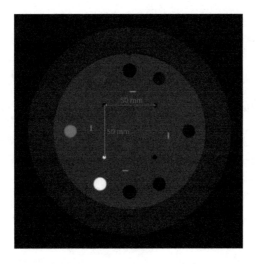

FIG. 50. Annotated computed tomography image of a phantom with known dimensions.

images, either the software tool or the system needs a recalibration. Contact the service engineer to assess the system.

5.2. QUALITY CONTROL TESTS FOR MEDICAL PHYSICISTS

All general tests performed by the radiographers should be revised and performed annually by the medical physicists.

5.2.1. Computed tomography number accuracy, image noise, image uniformity and image artefacts

5.2.1.1. Description and objective

The test consists of scanning a water filled test object (or a phantom containing uniform material) to ensure that the CT number and noise levels and the uniformity of CT images are within tolerances and that no image artefacts are visible [23, 24].

5.2.1.2. Equipment

The manufacturer's phantom, a commercial phantom or a simple round phantom made of water equivalent material is needed for this test.

5.2.1.3. Procedure

The test procedure is as follows:

(a) Use a CT acquisition protocol specified by the physicist at commissioning or the manufacturer's QC protocol.
(b) Centre the phantom in the tomographic plane and scan it using the specified protocol.
(c) Repeat the test for the most frequently used clinical protocols.
    Note: Pay special attention to slice thickness, as this has a great impact on noise.

5.2.1.4. Analysis and interpretation

The following steps are used for the analysis and interpretation of the results:

(a) Measure the CT number accuracy, image noise and uniformity; these parameters should be evaluated on the same images. Confirm the absence of artefacts on every image.

(b) Measure the CT number and image noise in a centrally placed circular ROI of appropriate diameter, as specified at commissioning (see Section 5.2.9).

(c) The size and position of the ROI are important. If the CT image quality is evaluated using QC software, use the specified ROI diameters. The following criteria are recommended:

　(i) For the measurement of the CT number, the diameter of the ROIs should be approximately 10% of the diameter of the image of the phantom (Fig. 49). These ROIs should be placed in the centre and at the 12, 3, 6 and 9 o'clock positions on the perimeter of the phantom, about 1 cm from the perimeter.

　(ii) For the measurement of noise, the diameter of the ROI should be approximately 40% of the diameter of the image of the phantom and the ROI should be placed in the centre (Fig. 49).

(d) The CT number is the measured mean value of CT numbers within a ROI, and the noise is the standard deviation of the CT numbers within a ROI.

(e) The uniformity is determined as the absolute difference between the CT numbers in the centrally placed ROI and those in each of the four ROIs on the perimeter (Fig. 49). Compare each of these four values with the given tolerance.

(f) Acquire several CT scans to obtain a mean noise value with greater precision.

(g) Visually inspect all images acquired during the noise test for image artefacts. For multidetector computed tomography scanners, this means examining all image slices in the acquisition, not just the central image slice.

5.2.1.5. Baselines and tolerances

The following baselines and tolerances apply:

(a) For the CT number accuracy: ±5 HU (for water) from the baseline value is acceptable, or within the tolerance provided by the manufacturer (if available). For other materials, refer to the specifications of the phantom.

(b) For image noise: ±25% from the baseline value is acceptable, or within the tolerance range stated by the manufacturer (if available).

(c) For the uniformity: ±10 HU (for water) is acceptable, or within the tolerance range provided by the manufacturer (if available).

(d) For artefacts: no artefacts with the potential to compromise diagnostic confidence should be visible.

5.2.1.6. Frequency

The tests are repeated every six months.

5.2.1.7. Corrective actions

If the measured values are out of the tolerance range, the images exhibit artefacts, or both, then first review the results and check the exposure parameters. The geometry of the set-up should not have a significant impact on the images. In case of artefacts, rotate the phantom to check the source of the artefact. If the problems persist, then it may be necessary to contact a service engineer.

5.2.2. Linearity

5.2.2.1. Description and objective

The aim of the test is to assess CT scanner performance for imaging materials of different absorbance. This is especially important for CT scanners used for radiotherapy treatment planning, as the representation of materials may impact treatment planning [23].

5.2.2.2. Equipment

An image quality test phantom with inserts of different attenuating materials is used for the test. Each attenuating material should have a known relative electron density and preferably a defined or designated HU value.

5.2.2.3. Procedure

The test procedure is as follows:

(a) Follow the CT acquisition protocol specified by the medical physicist at commissioning or the QC protocol provided by the manufacturer.

(b) Centre the phantom in the tomographic plane and scan it using the specified protocol.

(c)   Repeat the test for each kilovoltage value.

Note: The filtering of the image reconstruction kernel has an impact on the X ray attenuation and the representation of different materials. It is recommended to perform two image reconstructions, one with a kernel regularly used for soft tissue and one regularly used to image bones.

5.2.2.4.   Analysis and interpretation

Draw ROIs over the images and determine the mean HU value and standard deviation in each ROI of the inserts (Fig. 51).

5.2.2.5.   Baselines and tolerances

For water, the tolerance is ±4 HU compared to the baseline values; for other materials, it is −10 to 10 HU. For radiotherapy applications, typical values are usually provided by the manufacturer of the linearity measurement phantom; with reference to these values, the tolerance is −20 to 20 HU.

5.2.2.6.   Frequency

The test is repeated annually.

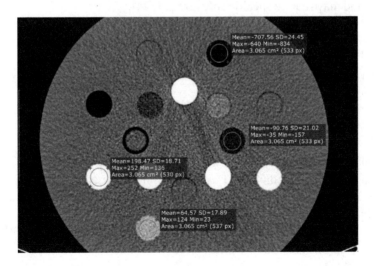

FIG. 51.   Phantom image with ROIs used for linearity measurements using a computed tomography quality assurance phantom made of materials with different absorbance.

5.2.2.7. Corrective actions

If the X ray attenuation significantly deviates from that determined as the baseline during commissioning, then review the results. Check the exposure parameters and image reconstruction used throughout the test, as related differences may lead to inconsistencies. After repeating the test, if the results are still outlying, contact the service engineer. If a significant deviation from the expected baseline occurs, then check the dosimetry of the CT system (Section 5.2.7). Review any software updates or changes to the image reconstruction algorithms used. If the CT machine otherwise provides the expected performance, then consult the service engineer about the results and establish a new baseline, if necessary.

5.2.3. Low contrast detail detectability

5.2.3.1. Description and objective

The aim of the test is to assess the response of the system to objects of low contrast differences and to verify that the low contrast performance of clinical protocols has not changed. This test requires a special image quality test object tailored for this assessment. Furthermore, many guidelines do not consider this test essential, as the information on issues that may arise from the imaging system are available from other tests. Objective image quality evaluation is difficult and may require specific software [23, 24].

5.2.3.2. Equipment

An image quality test phantom with low contrast detail objects is needed for this test (Fig. 52).

5.2.3.3. Procedure

Align the phantom with the laser positioners. Make an exposure for each kilovoltage that is used clinically. Ensure that clinically used image reconstruction parameters are selected and post-exposure reconstructions are done when necessary. Follow the manufacturer's specifications, if available.

5.2.3.4. Analysis and interpretation

The visibility of each contrast detail object should be evaluated either visually on a display device used for clinical evaluation of images or by using

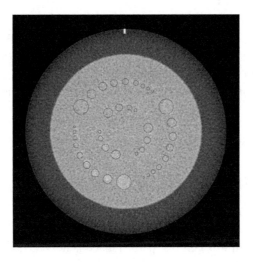

FIG. 52. Computed tomography image of phantom with low contrast detail insets marked with red circles.

ROIs drawn over each object and determining the low contrast detail detectability in a quantitative manner.

(a) Visual analysis:
   (i) View each series of images and identify the image that provides the best low contrast performance.
   (ii) Adjust the window width and level to optimize visibility of the low contrast targets (e.g. window width of 100 HU and window level of 100 HU).
   (iii) Record the size and/or contrast of the barely visible targets. Usually, it is convenient to use an indicator showing the visible objects per total number of objects.
(b) Numeric analysis:
   View each series and identify the image that provides the best low contrast performance. This mode of evaluation requires software based on model observers.

5.2.3.5. Baselines and tolerances

Appropriate performance criteria need to be established for the phantom used by the medical physicist during commissioning. The results and findings should be discussed with a team consisting of a radiologist, the medical physicist and a radiographer, in order to determine protocols that optimize dose and image quality. This should be based on the clinical needs for a specific protocol.

If the medical physicist selects the manufacturer's specifications as the tolerance, the scan technique should ideally be identical to the manufacturer's recommendation (including reconstructed image thickness and reconstruction kernel or filter).

5.2.3.6. Frequency

The test is repeated annually and after changes.

5.2.3.7. Corrective actions

If the measured low contrast performance is not satisfactory, then perform the following actions:

(a) Review the acquired results.
(b) Check the clinical scan protocol, the exposure parameters and the geometry set-up.
(c) Repeat the test after completely resetting the test set-up.
(d) Check whether one of the following situations causes the degradation:
  (i) The reconstructed image thickness is inappropriately thin;
  (ii) The mAs value is set inappropriately low;
  (iii) The pitch is too high for the clinical requirements, especially in systems that set the pitch and mAs value independently;
  (iv) The selected reconstruction algorithm is too sharp.

Degradation of the low contrast detail detectability indicates an overall decline in the performance of the imaging system. After checking the change of clinical protocols and set parameters used to test the equipment, the test should be repeated. If the deviation is still observable, then the manufacturer should be contacted and a further inspection should be carried out to investigate the root cause.

5.2.4. X ray beam width

5.2.4.1. Description and objective

The X ray beam width is a measure of the collimated beam width along the vertical axis. The objectives of this test are the following:

(a) To determine the accuracy of the collimator settings;
(b) To determine the extent of overbeaming (i.e. the difference between the image width and the beam width [23, 24]).

5.2.4.2. Equipment

A special X ray detector is recommended to be used, such as a computed radiography plate, a self-developing film or another similar film test tool. Some instrument manufacturers produce semiconductor detectors that can obtain the dose profile and measure the CT dose index (CTDI) with the appropriate corrections. Such tools are convenient, but not absolutely necessary. Alternatively, luminescent dosimeters may be used.

5.2.4.3. Procedure

The test procedure is as follows:

(a) Use a flat foam block to minimize scatter on the detector.
(b) Adjust the table height so that the detector is at the isocentre.
(c) If applicable, mark the isocentre on the detector (Fig. 53).
(d) Adjust the exposure parameters according to the response of the detector. If the reading is too low or not easily distinguishable, then raise the mAs value; otherwise, lower it.
(e) Scan the detector in axial mode using each beam width available.

5.2.4.4. Analysis and interpretation

Determine the beam profile from the obtained image and measure the full width at half-maximum.

5.2.4.5. Baselines and tolerances

See the manufacturer's specifications. If these are not available, then $\leq+3$ mm or $\leq+30\%$ of the total nominal collimated beam width, whichever is greater, is acceptable. The measured collimation should not be less than the nominal.

5.2.4.6. Frequency

The test is repeated annually and after changes.

FIG. 53. Self-developing film with isocentre position marked.

5.2.4.7. Corrective actions

If the tolerances are not met, review the acquired results and check the geometry of the set-up. Then, repeat the test after completely resetting the test set-up to check for errors. Contact the service engineer if the new measurement yields the same result.

If the beam is not wide enough to irradiate all of the detector elements, then this will probably impact clinical use of imaging significantly. If the X ray beam is larger than the given tolerance, then a large portion of the X ray beam is irradiating the patient but not contributing to imaging.

5.2.5. Reconstructed image slice width

5.2.5.1. Description and objective

This test ensures that the reconstructed image slice width corresponds to that selected on the CT scanner console. This test also confirms that the system performance does not change after repair, maintenance or service [23].

5.2.5.2. Equipment

This test uses a phantom that typically has thin metal inclined planes or other insets to provide information on slice width (Fig. 54). Depending on the phantom, it may be necessary to omit very thin slices, as the test object itself has an attenuation (e.g. 1 mm aluminium).

5.2.5.3. Procedure

According to the user manual of the phantom, obtain scans preferably in both helical and axial scan modes.

5.2.5.4. Analysis and interpretation

Perform the measurement according to the phantom manufacturer's specifications. Verify the phantom specifications and apply the appropriate corrections if needed.

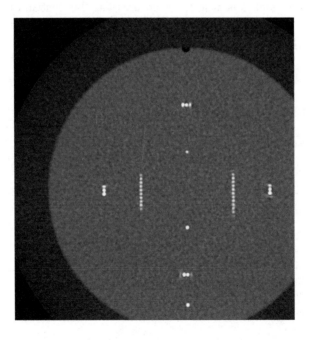

FIG. 54. Image of phantom with beads used to measure the reconstructed slice width.

5.2.5.5. Baselines and tolerances

Table 5 provides the suggested tolerances.

5.2.5.6. Frequency

The test is repeated annually and after changes.

5.2.5.7. Corrective actions

Repeat the test after completely resetting the test set-up to confirm the error. If a deviation is observed during this test, then check the X ray beam width (Section 5.3.9). Contact the service engineer if the new measurement yields the same result. Consult with the staff using the CT machine about whether this issue impacts clinical diagnosis, and take action accordingly.

5.2.6. Spatial resolution

5.2.6.1. Description and objective

The purpose of this test is to ensure that the spatial resolution of a reconstructed image complies with the manufacturer's standards. This test is optional [23].

TABLE 5. TOLERANCE FOR IMAGED SLICE WIDTH

Nominal slice width (mm)	Acceptable
≤1	Less than nominal + 0.5 mm[a]
>1 and ≤2	±50%
>2	±1 mm

[a] Thin slices made in the axial scanning mode may have a much larger value than expected owing to the thickness of the ramp (e.g. 3 mm rather than 1 mm for a setting of 1 mm). In this case, refer to the user manual of the phantom and apply the necessary correction factors.

5.2.6.2. Equipment

The following equipment is needed for the test:

(a) MTF measurements are performed with objects that have a high contrast with respect to a uniform background or with a phantom that incorporates high contrast targets of known resolution. This is usually a bead or a similar object designed to determine the MTF.
(b) Software for MTF evaluation.
(c) Alternatively, instead of measuring the MTF, a phantom with a line resolution test pattern may be used for subjective visual evaluation. With this method, only the limiting spatial resolution can be determined.

5.2.6.3. Procedure

The test procedure is as follows:

(a) Centre the phantom on the tomographic plane.
(b) Align the test object parallel to the image plane.
    Note: Alignment is important if a bar pattern or rods are used (Fig. 55).
(c) Select a relevant scan protocol or one specified at commissioning and is appropriate for the phantom in use.
(d) Scan the phantom.
(e) Repeat for additional scan protocols if needed.

5.2.6.4. Analysis and interpretation

The following procedures are applied:

(a) For resolution patterns (e.g. bars, rods), visual analysis is appropriate;
(b) For MTF evaluation, the image needs to be evaluated with the appropriate analysis software.

5.2.6.5. Baselines and tolerances

The baselines and tolerances provided in the CT manufacturer's specifications are used.

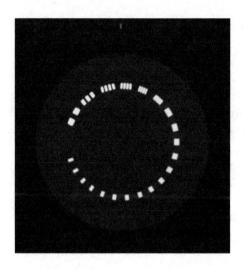

FIG. 55. Bar pattern phantom image with the alignment inset shown at the top.

5.2.6.6. Frequency

The test is performed during acceptance and after changes that affect imaging.

5.2.6.7. Corrective actions

If tolerances are not met, review the acquired results and check the geometry of the set-up. Repeat the test after completely resetting the test set-up to confirm the error. If the X ray parameters are appropriately set, perform a recalibration or ask the service engineer to perform one. Degradation of the MTF may have several causes, such as blurring of the focal spot due to ageing of the X ray tube, loss of sensitivity of the detector elements or change in the gain of the amplifiers of the detector elements. The latter may be confirmed by a measurement of homogeneity and noise (Section 5.3.6). The service engineer should be contacted if the new measurement yields the same result.

5.2.7. Computed tomography dosimetry

5.2.7.1. Description and objective

The CTDI measurement ensures the constancy of the X ray tube output and assists in the optimization of patient exposure. The objectives of this test are the following:

(a) To review and update the volumetric CTDI, C_{vol} (in other publications: $CTDI_{vol}$), and the dose–length product values for selected protocols representing routine examinations;
(b) To measure a small set of CTDI values measured in air, C_{air} (in other publications: $CTDI_{air}$), to confirm output consistency;
(c) To review, for selected scan parameters, the accuracy of the displayed C_{vol} and kerma–length product values on the scanner console, if applicable [23–27].

5.2.7.2. Equipment

The following equipment is required for the test:

(a) A dosimeter with a calibrated pencil ionization chamber, compliant with the relevant IEC standard [25];
(b) An electrometer;
(c) Standard head and body CT dosimetry phantoms (see Fig. 56);
(d) A chamber stand for a free-in-air measurement;
(e) Devices to stabilize and secure the phantom.

FIG. 56. Set-up for the measurement of dose in a standard CTDI PMMA body phantom (left) and a free-in-air measurement (right).

5.2.7.3. Procedure

The following procedures are used:

(a) Measurement of C_{vol}:
 (i) Align the body phantom.
 (ii) Place the ionization chamber in the central hole of the phantom and use PMMA plugs to fill unused holes.
 (iii) Check the position using the SPR. This test is not sensitive to positioning, so a tilting of about 5° is acceptable (Fig. 56).
 (iv) Select an axial scanning protocol specified during commissioning for the given phantom type. Do not apply any dose reduction technique or the AEC. Some scanners may over-rotate unless the rotation is restricted to 360°; this should be considered during the evaluation.
 Note: It is recommended for head and body protocols that every kilovoltage and at least three different mAs values (e.g. 50, 100 and 300 mA · s) are checked during commissioning and that all nominal beam widths are also selected during testing.
 (v) Make an exposure with the selected parameters.
 (vi) Record the measured dosimeter reading, which (for most measuring instruments) is the integral of the dose profile for 100 mm (C_{100}) and apply correction factors if necessary.
 (vii) Record the C_{vol} value displayed on the acquisition workstation monitor for that particular scan, if available. This value is usually updated after the exposure.
 (viii) Reposition the ionization chamber into one of the peripheral holes of the phantom.
 (ix) Repeat steps (v) through (viii) after moving the ionization chamber into the next hole. A total of five exposures (one in the centre and four in the periphery) should be made for each set of scan parameters. To reduce the uncertainty of the measurement, each measurement can be repeated and the resulting values averaged.
 (x) Repeat the whole procedure with the head phantom seated into the headrest, using the applicable axial protocol.
(b) Measurement of C_{air}:
 (i) Align the ionization chamber in the centre of rotation and check the position using the SPR. A position accuracy of ±1 cm is acceptable.
 (ii) Verify the centring using the lasers or by taking a single axial slice.
 (iii) Select approximately 120 kV and 100 mA · s for the clinical protocol typically used (e.g. CT head) with the largest X ray beam collimation, and scan the ionization chamber in axial mode.

(iv) Record the measured C_{100} value and apply correction factors if necessary.

(v) Repeat the measurements for other X ray beam collimations (e.g. using the lowest and highest kilovoltage at the reference X ray beam collimation widths). The suggested reference collimation beam width is 20 mm or the largest possible collimation width below 20 mm.

(vi) Record the relevant exposure parameters including scan length for the calculation of kerma–length product (dose–length product).

If the X ray beam is wider than 40 mm, see Ref. [27].

5.2.7.4. Analysis and interpretation

Measurements should be compared with the baseline values and with the dose index values displayed on the acquisition workstation monitor.

(a) Calculate C_{vol} for the central and peripheral measurements from the corrected dosimeter readings. Repeat this for each phantom and protocol used.

(b) Calculate the kerma–length product from C_{vol}.
Note: For a single axial rotation, the weighted CTDI (C_w) and C_{vol} are identical.

(c) Compare C_{vol} with the baseline and displayed dose index values.

(d) Measure C_{air} for all beam collimations.

(e) Calculate C_{air} from the dosimeter readings for all X ray beam collimations and kilovoltages used.

(f) Compare these values with the results of commissioning or previous constancy tests.

5.2.7.5. Baselines and tolerances

The following apply:

(a) For C_{vol}, ±20% difference between the baseline or displayed values and the measured values is acceptable.

(b) C_{air} should confirm the tolerances stated by the manufacturer. If no such tolerance is available, then ±20% deviation from the baseline is acceptable.

5.2.7.6. Frequency

The test is repeated annually or after changes that affect dosimetry.

5.2.7.7. Corrective actions

Non-conformance of the displayed dose indicators to the measured values should be investigated. Consistent errors across imaging protocols and settings may be due to wrong calculation factors implemented by the manufacturer. Differences below the tolerance can be corrected for; however, larger deviations should be investigated. Inconsistencies or issues in the linearity of the dose with the mAs value indicate a problem with X ray generation. If only the dose indication fails, the system may be used, but the problem should be repaired when possible.

6. TESTS FOR ALL
DIAGNOSTIC RADIOLOGY SYSTEMS

6.1. QUALITY CONTROL TESTS FOR RADIOGRAPHERS

6.1.1. Routine check of image display

6.1.1.1. Description and objective

Correct set-up of the display monitors is essential to achieve a good diagnostic outcome. The objective of this test is to confirm that the image displays reproduce all of the grey scale information in the image accurately [23].

6.1.1.2. Equipment

The QC test pattern recommended by the American Association of Physicists in Medicine Task Group 18 [28], AAPM TG18, is used for this test (Fig. 57).

6.1.1.3. Procedure

The test procedure is as follows:

(a) Display the AAPM TG18 test pattern on the image display, as specified by the medical physicist at commissioning.
(b) Set the window width and level at the values defined during commissioning.
(c) With the room lights set at the illumination level that is normally used, view the images on the image display.

(d) Carry out the test for each image display associated with the X ray system (e.g. operator's console, radiologist's workstation) to ensure that the displays produce images of similar quality.

6.1.1.4. Analysis and interpretation

The following procedure is used:

(a) Evaluate each image carefully (see Fig. 31). Determine whether it is possible to see brightness or density differences between adjacent steps of the step wedge. Observe the visibility of the 5% and 95% inset patches.
(b) Note the 5% and 95% patches inset inside the 0% and 100% patches, respectively. These should be visible on the image displays.

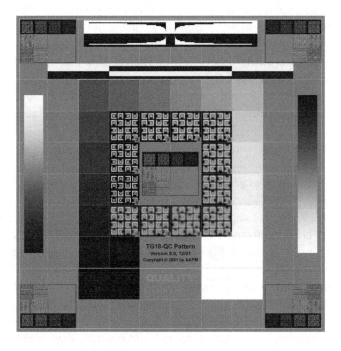

FIG. 57. AAPM TG18 QC test pattern. The 0% patch usually cannot be visualized on a printed (reflective) image such as this figure, but it should be visible on displays. Image reproduced with permission from Ref. [28].

6.1.1.5. Baselines and tolerances

Each of the steps of the step wedge should be visible and distinguishable. The tolerance for object visibility should be as for the patch visibility and the visibility of the insets (5% and 95% contrast), acceptable findings are that adjacent steps are distinguishable and all low contrast inset patches are visible.

6.1.1.6. Frequency

The test is repeated monthly.

6.1.1.7. Corrective actions

If loss of visibility is detected, then the following aspects need to be considered:

(a) The image display may have changed. Ensure that the display is set correctly.
(b) Loss of visibility may be caused by a change in room illumination levels. When room lighting increases, subtle differences are more difficult to discern. This is easily observed by inspecting the low contrast (5%) inset patches. Adjust the room lighting.
(c) If the above steps do not solve the problem, investigate more thoroughly with the medical physicist to decide on the appropriate corrective actions.

6.1.2. Reject rate assessment

6.1.2.1. Description and objective

With the advent of digital imaging, assessment of the reject rate is not performed as consistently as when each rejected film was strictly accounted for. However, with the use of the appropriate software it is easy to retrieve and analyse rejected images.

Digital imaging is less prone to producing over- or underexposed images than films were. Nonetheless, errors are made during imaging procedures, leading to repeated acquisitions in order to obtain an appropriate image for diagnostic purposes. If these errors are observed and documented, they could be avoided or at least their frequency could be decreased. The ease of repeating an imaging procedure is both a merit and a drawback of digital systems, as they enable X ray exposure of patients with the push of a button. Learning to eliminate rejected images helps to reduce patient exposure [29].

6.1.2.2. Equipment

The following equipment is required for this test:

(a) Rejection charts and/or logs or remote data collection software (if available);
(b) Appropriate charts or a standardized list of reasons for rejection to enable follow-up.

6.1.2.3. Procedure

The test procedure is as follows:

(a) Whenever image acquisition is repeated or an image is rejected for any reason, it should be documented. Most conveniently, the radiology information system or the software of the equipment could be used for this purpose if such a feature is implemented. A Digital Imaging and Communications in Medicine (DICOM) structured report or a similar record could be used to report reasons known immediately or to indicate that further investigation is necessary. This document includes the following standard reasons for rejection:
  (i) Positioning (e.g. rotation, anatomy cut-off, incorrect projection, incorrect marker);
  (ii) Exposure error (e.g. overexposure, underexposure);
  (iii) Grid error (e.g. cut-off, decentring, no grid, grid lines);
  (iv) System error;
  (v) Artefacts (e.g. foreign objects, contrast media, imaging system);
  (vi) Patient motion or other movement error;
  (vii) Test images;
  (viii) Study cancelled;
  (ix) Other.
(b) In addition, the following information should be noted:
  (i) Acquisition station, digitizer;
  (ii) Accession number;
  (iii) Examination date and time;
  (iv) Body part;
  (v) View;
  (vi) Exposure indicator (exposure index);
  (vii) Reject category;
  (viii) Identifier(s) of the radiographer;
  (ix) Technique factors;
  (x) Thumbnail.

(c)   After collecting the information for a pre-set time period (one to three months), a review should occur.

6.1.2.4.   Analysis and interpretation

After the predetermined time, a review of the collected data should occur. The reject rate by itself is insufficient for further investigation. However, with the appropriate statistics, the reasons and the trends can be analysed.

6.1.2.5.   Baselines and tolerances

It is recommended to adopt a low rejection rate and that the radiology department use its own figures as benchmarks. A 4% to 8% rejection rate is reported in imaging departments using digital imaging modalities, while 3 to 5% is achievable in paediatric departments.

The target should be set locally by a team from the department, considering that rates that are too low lead to the acceptance of poor quality images, while other factors (e.g. employment of trainees) also affect the possible target rate.

6.1.2.6.   Frequency

It is desirable that the assessment be performed monthly or at least every three months.

6.1.2.7.   Corrective actions

After reviewing the statistics for the most significant component of the reject rate, the common reason should be determined, and an appropriate corrective action should be introduced by the imaging team.

6.1.3.   Visual inspection of the workplace

6.1.3.1.   Description and objective

This routine test is needed to verify the mechanical and electrical operation of the X ray system and to ensure that the image acquisition information is correct. Originally, this test was developed for mammography; however, most of this test is applicable to other modalities [18, 23].

6.1.3.2. Equipment

A thermometer is needed for this test.

6.1.3.3. Procedure

The test procedure is as follows:

(a) Measure the temperature in the acquisition room;
(b) Visually inspect the unit for loose parts;
(c) Check the cleanliness of the surfaces that come into contact with the patients (e.g. bucky, couch) and the overall integrity of the system;
(d) Check the compression paddles for cracks;
(e) Ensure that a faceguard is present on mammography units and does not have any damages;
(f) Check couches and patient support for damage (which may cause artefacts);
(g) Check cables for any bending, breaks or knots;
(h) Verify that no cables or hoses are under heavy objects;
(i) Check that the angulation indicator is working and indicates the correct value (if applicable);
(j) Verify that interlocks work as designated and are accessible;
(k) Verify smooth movement of any moving parts, including a gantry or arm holding the imaging system;
(l) Verify the proper function of all auxiliary controls and indicators, such as switches, pedals, panels, indicator lights and any measuring device;
(m) Check the laser indicators and the light field (if applicable);
(n) Verify that a technique chart is available if the system does not have an AEC;
(o) Check the annotations, the facility information, the time and date on the workstation by opening a recently made image (if applicable);
(p) Verify that films have the appropriate annotation (if applicable);
(q) Verify that the compression on a mammography unit releases when the power fails;
(r) Confirm the integrity of the operator shield or the windows between the operator's room and the acquisition room. It should not have any objects (clutter) impairing free visibility;
(s) Ensure availability of the cleaning solution recommended by the system's manufacturer;
(t) Verify any other functions that are specified for regular monitoring by the equipment's manufacturer;
(u) Ensure that written radiation safety procedures are available and up to date;
(v) Ensure that the examination room is clean and that the door can be closed;

(w) Check that lead aprons and other required shielding and portable barriers are available (if necessary), have no defects and are clean;

(x) Ensure that emergency equipment and phone numbers are present;

(y) Check that any further facility specific items are present and functioning.

6.1.3.4. *Analysis and interpretation*

It is advisable to use a checklist to keep track of possible changes in the environment.

6.1.3.5. *Baselines and tolerances*

The following requirements apply:

(a) Room temperature should be in the range recommended by the manufacturer and comfortable for patient care.

(b) All mechanical and electrical items should be in a satisfactory state of operation.

(c) The time and date, as well as the facility identification, need to be correctly displayed in the image annotation on the interpretation workstation.

6.1.3.6. *Frequency*

The test is repeated monthly or after any service, maintenance and software upgrades.

6.1.3.7. *Corrective actions*

Possible corrective actions and their suggested time frames for several of the above crucial aspects are listed in Table 6. In addition, the following actions need to be taken:

(a) If the room temperature is not in the range recommended by the manufacturer, the custodian or the local maintenance technician should be contacted.

(b) The appropriate service personnel should repair items that are hazardous, inoperative or out of alignment, or items that operate improperly, before any further patients undergo imaging.

(c) Items missing from the room should be replaced immediately.

(d) Malfunctioning equipment requires immediate repair.

(e) Before performing any further imaging, deficiencies should be corrected by the appropriate personnel.

TABLE 6. SUGGESTED TIME FRAME FOR CORRECTIVE ACTIONS

Immediate action	Action within 30 days
Room temperature not controlled	Angulation indicator not functioning
Loose parts, paddles damaged or bucky not clean	Gantry motion not smooth
Hoses or cables kinked or damaged	Faulty panel switches, pedals, indicator lights and meters
Interlocks faulty	Laser or light inoperative
Time, date or facility identification incorrect or not present in the image	Current technique chart not posted
Cleaning solution not available	Operator radiation shield damaged
Automatic and/or manual compression release not working (mammography only)	

6.2. QUALITY CONTROL TESTS FOR MEDICAL PHYSICISTS

6.2.1. Image display systems

6.2.1.1. Description and objective

The quality of the image display and the associated viewing conditions for the acquisition console and workstations are tested to ensure the following:

(a) The image is represented without distortions and in the appropriate size;
(b) The room illuminance and display reflection and the luminance response are set properly;
(c) The display resolution, low and high contrast and display noise are within tolerances.

An updated guide for display QA was published by the AAPM [30] during the preparation of this handbook. It is suggested to consider the most recent version for the revision of the local QA programme.

For printer QC testing, further recommendations are provided in Refs [23, 24, 31].

6.2.1.2. Equipment

The following equipment is required:

(a) A photometer capable of measuring a small area (5–10 mm diameter) of luminance on displays, to measure both luminance and illuminance;
(b) An AAPM TG18 QC test pattern.

6.2.1.3. Procedure

The test procedure is as follows:

(a) Display the AAPM TG18-QC test pattern on the image display.
(b) Set the window width and level to the specified values for the pattern. Do not set it by eye.
(c) With the room lights set at the illumination level normally used, examine the test image on the display using a calibrated photometer. Measure the luminance of each patch. The luminance levels of the 90% and 40% patches, along with those of the 40% and 10% patches, are indicative. Their difference should be determined after the measurement.
(d) Determine the illuminance of the room.
   Note: Each photometer uses different sensors to measure luminance and illuminance. Ensure that the instrument's setting is appropriate for the given sensor.

6.2.1.4. Analysis and interpretation

The following procedure is used:

(a) Evaluate the images. If film printers are used, check the images on both the display and the film.
(b) Carefully check each step wedge and record whether significant differences are noticeable between adjacent steps.
(c) Ensure that the 5% and 95% inset patches are visible.
(d) Check the contrast by observing the low and high contrast patterns.
(e) Check all lines for geometric distortion.

TABLE 7. TOLERANCES FOR IMAGE DISPLAY TEST

Test quantity	Acceptable
Visibility of step wedge patches; 5% and 95% inset patches	The brightness or density differences between adjacent steps of the step wedge should be visible. Both the 5% and 95% inset patches should be visible.
Display maximum luminance levels[a]	Primary workstation display: $\geq$350 cd/m^2 Secondary display: $\geq$250 cd/m^2 For mammography: $\geq$420 cd/m^2
Room illuminance levels	15–50 lux

[a] Typical values for LCD displays.

Note: 'Primary display' refers to the workstation used by the radiologist for primary interpretation. 'Secondary display' refers to displays used by other physicians for reviewing images.

6.2.1.5. Baselines and tolerances

Table 7 presents suggested tolerances for object visibility, luminance level and room illuminance.

6.2.1.6. Frequency

The test is repeated annually.

6.2.1.7. Corrective actions

The following corrective actions are applicable:

(a) Lower the room illumination levels to below 15–50 lux, as higher values could impact the visibility of the 5% and 95% inset patches.

(b) Adjust the brightness and contrast settings of the display if the room lighting is set correctly.

(c) Re-adjust the display in an automatic manner using the appropriate software. If that function is not available and the brightness and contrast settings are independent, then the following procedure could be performed while displaying the test image:

(i) Set the display brightness and contrast levels to zero.

(ii) Adjust the brightness setting up to the point where the background becomes barely visible.

(iii) Change the contrast so that the 95% patch is visible.

(iv) Increase the contrast until the alphanumeric symbols on the image show distortion; then, revert the contrast to just before distortion appears.

(v) If the brightness and contrast settings are not independent of each other, then set a lower brightness setting to create a darker background, and readjust the contrast. Ensure that the 5% patch stays visible when adjusting the brightness.

(d) The 5% and 95% patches should be visible simultaneously. If it is not possible to set the display to achieve this, contact the service engineer.

6.3. DOCUMENTATION

It is important to emphasize that without keeping records of the results, any QA programme is meaningless. These records show the activities conducted and serve as a reference to follow up on any changes and issues that may affect patient care. QA is helpful not only for the team responsible for quality but also for the service engineer to easily identify performance issues. Record keeping of the QC tests outlined in this handbook also provides evidence for regulatory purposes, as inspectors could check records and confirm compliance with the regulatory requirements.

When QC tests are performed, it is advised that the following information is recorded in a clear and logical manner:

(a) General information:
   (i) Identification of the subject (e.g. 'daily QC test of the XY mammography unit');
   (ii) Time and date of the test;
   (iii) Place where the test was carried out and place where the system is used, if different (e.g. name, address, building, room);
   (iv) Identification of the organization or personnel performing the test (i.e. name and contact details);
   (v) Identification of the licensee, operator or user of the equipment (e.g. name, contact details, position);
   (vi) Identification of the equipment (e.g. manufacturer, type, serial number of the whole system or its main components; if imaging requires further components, then the information related to those is also recorded);

(vii) Clear page numbering if the report consists of several pages (e.g. current/total pages).
(b) Instruments and test objects:
  (i) Identification of the measuring devices and test objects (e.g. manufacturer, type, serial number, firmware version if applicable);
  (ii) Information on the calibration of the instruments;
  (iii) Information on the measurement uncertainty (may be calculated for each individual measurement);
  (iv) Settings, modes of operation and any other factor that may affect the measurement (once or for each measurement), including correction factors.
(c) Environmental conditions and use information (if relevant):
  (i) Temperature and pressure;
  (ii) Floor plan or a sketch of the room.
(d) Technical information on the equipment:
  (i) Specifications (e.g. tube voltage range, AEC modes of operation, software version);
  (ii) Any reported failure of the equipment or major changes affecting any parts.
(e) Tests performed:
  (i) List of tests performed;
  (ii) Scope or brief description of the test, or references to the guidelines used for the test;
  (iii) Exposure parameters or modes of operation;
  (iv) Results (i.e. readings, corrected values and factors used to obtain the results);
  (v) Evaluation of the results.
(f) Conclusions drawn from the tests:
  Short summary and list of corrective actions with their corresponding deadlines and the personnel responsible to carry these out.
(g) Signatures, stamps or any other verification.

6.4. DOSE MANAGEMENT SOFTWARE

Dose management software (DMS) has recently been introduced to diagnostic radiology to facilitate QA and patient dose management [32]. Apart from being an essential tool for improving a QA programme and ensuring regulatory compliance, DMS can be used to collect, monitor and evaluate patient demographics and technical information, including dose data from various imaging modalities [32]. Depending on the type of DMS, apart from collecting information on quantities relevant to patient dose, it can also extract other data to further enhance patient care and assist with continuous improvement of practice quality. The features and capabilities of DMS determine its utility; not all types of software are applicable to every situation. Care should be taken when selecting DMS for a given department to suit local needs. DMS comes in many forms and it may be developed by international commercial companies or be available for free or as an open source solution. DMS is efficient at improving everyday clinical practice and is particularly important as a quality management tool. It provides improved automated ways to collect, archive, analyse and report technical data compared with manual or semiautomated data collection methods (Fig. 58).

Whatever software is selected, dose indices should be verified first at the X ray system level and then together with all other relevant information to ensure that these are accurately transferred to the DMS platform. This should be realized by the clinically qualified medical physicist [32]. Reliable and accurate use of DMS in the everyday clinical routine can be accomplished only after a comprehensive QA programme for radiological equipment is established and data verification is regularly performed.

FIG. 58. Summary report of mammographic examinations using DMS.

7. FILM–SCREEN SYSTEMS

The following tests serve as constancy tests for film–screen systems both for radiography and mammography systems. While these tests are relatively simple and may be performed by a trained and equipped medical radiation technologist (radiographer), it is recommended that a medical physicist periodically check the results, their interpretation and the corrective actions taken.

7.1. QUALITY CONTROL TESTS FOR RADIOGRAPHERS

7.1.1. Temperature of the film processing system

7.1.1.1. Description and objective

The temperature of the automatic or manual film processor should be evaluated. This temperature has an impact on film development and may lead to corrupting otherwise optimally exposed films, resulting in the repetition of the imaging procedure [4, 22, 33].

7.1.1.2. Equipment

Digital thermometer, preferably a calibrated one. Thermometers with mercury or other liquids should not be used, as possible faults could spoil the solution.

7.1.1.3. Procedure

The test procedure is as follows:

(a) After the stabilization of the film processor (allow sufficient time — for example, one hour — for the temperature to stabilize), perform the measurement.
(b) Disconnect the film processor from the mains supply to avoid electrical shocks.
(c) Remove the lid or cover of the developer liquid tank and submerge the thermometer into the solution. Depending on the make and type of the thermometer, wait until the reading stabilizes. Try to always measure at the same depth and position in the tank, as the solution may not be homogeneous.

Note: Always use the appropriate personal protective equipment when dealing with hazardous chemicals.

(d)   Record the reading of the thermometer.

7.1.1.4.   Analysis and interpretation

Depending on the films used, the temperature should be around a baseline value. Small fluctuations may occur, but the temperature should be relatively constant. Compare the current reading with the baseline or the manufacturer's recommendations.

7.1.1.5.   Baselines and tolerances

The recommended deviation is ±1.0°C.

7.1.1.6.   Frequency

The test is repeated daily.

7.1.1.7.   Corrective actions

Suspend use of the system should until the system is repaired. The service engineer should be contacted if the measured temperature is outside the tolerance range.

7.1.2.   Base and fog, sensitivity and contrast

7.1.2.1.   Description and objective

Sensitometry is the measurement of the film response to exposure and processing. A dedicated sensitometer consists of a uniform and constant light source and a sensitometric strip. This test provides information on the quality of the film and the development process, revealing relevant errors independently of the X ray system [22].

7.1.2.2.   Equipment

The following equipment is needed for the test:

(a)   Sensitometer with film strip (with 11 or 21 steps);
(b)   Densitometer.

7.1.2.3. Procedure

The following procedure is used:

(a) Expose the film in the sensitometer using the manufacturer's instructions. **Note:** Films sensitive to blue and green light should be checked with the corresponding light source.
(b) Process the film using the regular procedure.
(c) Use the densitometer to measure the optical densities obtained.

7.1.2.4. Analysis and interpretation

It is not necessary to note all the measured values for each step on a daily basis. At a minimum, the optical density values associated with base and fog, the speed point or speed index (OD = 1 above the base and fog value) and the contrast index should be determined. The contrast index is the difference between the optical densities of the step closest to OD = 0.25 above base and fog and that at OD = 2.0 above base and fog. The steps used for the speed index and contrast index measurement should be determined during the establishment of the baseline values.

7.1.2.5. Baselines and tolerances

Baseline values should be established if the film processor is operated under optimal circumstances for a short period. After the commissioning of the film processor (or if the film processor is cleaned, has fresh chemicals and a temperature specified by the manufacturer), repeat the whole process during the course of one week and determine the base and fog, contrast index and speed index. The average of the values obtained will serve as a baseline for the constancy tests. The same box of films should be used for the establishment of the baseline values.

Depending on the use of the film (radiography or mammography), different tolerances of the optical density are used. For general radiography, OD = ±0.15 of the baseline value can be used as a suspension level. With mammographic films, more consistent values can be achieved, so OD = ±0.10 may be used. For base and fog, an optical density limit of OD = +0.03 is appropriate.

7.1.2.6. Frequency

The test is repeated daily (after the temperature measurement).

7.1.2.7. Corrective actions

Ensure that the replenishment system is working correctly or, in the case of manual processing, try to repeat the test. The service engineer may be contacted for troubleshooting.

7.1.3. pH measurement of the fixer solution

7.1.3.1. Description and objective

The fixer solution is important for film processing, as it stops further development of the latent image. If the fixer is too acidic or too close to neutral pH, then film processing may stop too soon or too late, and the image will not have the achievable contrast characteristics. As with the previously described issues of the film processor, this could lead to corrupted images and unnecessary repetition of the imaging procedure [4, 33].

7.1.3.2. Equipment

The following equipment is required for the test:

(a) pH indicator strip for the 4.0–5.0 pH range;
(b) Comparison colour chart.

A dedicated instrument may be used (if calibrated), but reproducible results that are fit for the purpose can be achieved with the pH indicator method.

7.1.3.3. Procedure

The test procedure is as follows:

(a) Turn off the film processor after letting it warm up (e.g. 30 min).
(b) Remove the lid or cover from the fixer solution tank.
    Note: Always use the appropriate personal protective equipment when dealing with hazardous chemicals.
(c) Submerge the pH indicator strip for five seconds in the fixer solution.
(d) Remove the paper and dry it.
(e) Compare the colour of the strip with the comparison chart.

7.1.3.4. Analysis and interpretation

The result may indicate the change of the chemical status of the solution, which may impact image quality. The typical pH range of fixer liquids is 4.0–4.6; however, the manufacturer may specify a different value.

7.1.3.5. Baselines and tolerances

Follow the manufacturer's specifications for the pH value of the fixer solution.

7.1.3.6. Frequency

The test is repeated monthly.

7.1.3.7. Corrective actions

If necessary, renew the solution, as fixer needs continuous replenishment. Check whether the tank needs a thorough cleaning. Check other causes, as acidity is only one factor affecting the properties of the solution.

7.1.4. Maintenance and visual inspection of cassettes

The cassettes used for imaging should be inspected on a regular basis. Errors may be obvious during their daily use, but to prevent their occurrence, regular checks should be carried out. Cassettes are especially prone to errors, as they have moving parts (latches and hinges) and may be dropped during their daily use [4, 33].

7.1.4.1. Description and objective

The purpose of the test is to determine the status of cassettes and screens used for imaging.

7.1.4.2. Equipment

The following equipment is required for the test:

(a) Permanent marker pens;
(b) Labels;
(c) Cleaning solution;

(d) Padding.

7.1.4.3. Procedure

The test procedure is as follows:

(a) Clean the exterior of the empty cassettes with a dry cloth and then with a damp one; use alcohol or soapy water to dampen the cloth. Remove and secure screens separately to avoid spreading the cleaning solution on them.
(b) Visually inspect the cassettes, their hinges and locks or latches. Most usually the corners may be broken or may have cracks.
(c) Check for proper labelling of the screens inserted in the cassette. Always use the type of screen most appropriate for the films used for imaging (e.g. do not combine blue light emitting screens with green light sensitive films and vice versa).

7.1.4.4. Analysis and interpretation

If the cassette has defects (on the locks or hinges) or cracks, then their light-tightness should be evaluated.

7.1.4.5. Baselines and tolerances

Ensure that cassettes are light-tight and do not have any defects (see Fig. 59).

FIG. 59. Cassette on which the latch was stuck and pried open by force.

7.1.4.6. Frequency

Repeat the test every six months or when it seems appropriate.

7.1.4.7. Corrective actions

The screen type label should be reattached if it cannot be read properly. The screen and the cassette should bear the same labels. Repair any defects. Replace the cassette if necessary. Cassettes that have lost their light-tightness should be excluded from service.

7.1.5. Maintenance of screens

7.1.5.1. Description and objective

The aim of this test is to ensure that the screens are clean and do not have any artefacts that may impact diagnosis. Dirty or imperfect film–screen contact may produce artefacts on the image [4, 22, 33].

7.1.5.2. Equipment

The following equipment is required for the test:

(a) Non-alkaline soap solution with a composition and concentration approved by the screen manufacturer;
(b) Lint-free cloth;
(c) Brush.

7.1.5.3. Procedure

The test procedure is as follows:

(a) Open the cassette and inspect the screen under good lighting conditions.
(b) Look for defects on the surface of the screen.
(c) If there are suspected defects, such as visible scratches or similar imperfections, make a test exposure to determine the extent and size of the defects.
(d) When cleaning is performed, clean the intensifying screens with a soap solution and dry it with a cloth, leaving the cassette partially open in a vertical position.

Note: In order to spare films, it is possible to do the cleaning before a test exposure.

7.1.5.4.   Analysis and interpretation

Artefacts caused by defects of the screen may have an impact on the clinical diagnosis; a visual inspection or a test exposure may reveal such defects.

7.1.5.5.   Baselines and tolerances

No artefacts compromising clinical diagnosis should be visible (see Fig. 60).

7.1.5.6.   Frequency

Repeat the test weekly or when artefacts are suspected.

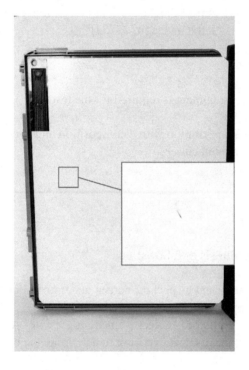

FIG. 60. Punctured screen.

Corrective actions

Clean the screens gently. If the screens still have serious faults that may have an impact on clinical diagnosis, then they should be rejected and removed from service.

7.1.6. Light-tightness of cassettes

7.1.6.1. Description and objective

Fogging of unexposed film clearly indicates a damaged cassette. The cassettes should shield the unexposed film against visible light [4, 33].

7.1.6.2. Equipment

The following equipment is needed for the test:

(a) Film viewer box or other strong light source;
(b) Densitometer (optional).

7.1.6.3. Procedure

The test procedure is as follows:

(a) Load a film into the cassette.
(b) Place the cassette on the film viewer box or around a strong light source to expose it.
(c) Leave the cassette exposed for 10 min on each side.
(d) Process the film in the regular manner.

7.1.6.4. Analysis and interpretation

Assessment of light-tightness is carried out by visually inspecting the processed film to determine whether it contains foggy regions (see Fig. 61) or a densitometer could be used.

7.1.6.5. Baselines and tolerances

No fogging of the film should be observed.

FIG. 61. Cassette that deteriorated over the years (inset), resulting in foggy images (main).

7.1.6.6. Frequency

The test is repeated every six months and as necessary.

7.1.6.7. Corrective actions

Reject the faulty cassette. It may be repaired, but light-tightness should be checked after repair.

7.1.7. Film–screen contact

7.1.7.1. Description and objective

The purpose of the inspection is to determine whether artefacts arise from bad film–screen contact [4, 33].

7.1.7.2. Equipment

A test object containing a fine metallic mesh is used for this test. A perforated metal sheet or a test object recommended by the film manufacturer could be used; if no specific test objects are available, then a metallic insect screen could be used as well.

The test procedure is as follows:

(a) Load a film into the cassette.
(b) Place the mesh or grid on top of the cassette.
(c) Centre the X ray tube at about 100 cm from the cassette; then open the collimator to let the radiation field cover the whole cassette.
(d) Set up an exposure that will result in a net optical density of OD = 1.0–2.0 or a value recommended by the test object's manufacturer.
(e) Process the film.

7.1.7.4. Analysis and interpretation

Place the processed film on a viewer box. The mesh should have high contrast on the image. If the image is blurry somewhere or if the mesh is not imaged properly, then the film–screen contact has defects. By observing the developed film from a distance, the faulty area will be seen as a region with higher exposure.

7.1.7.5. Baselines and tolerances

No artefacts should be observable.

7.1.7.6. Frequency

The test is repeated every six months.

7.1.7.7. Corrective actions

A bad film–screen contact may be due to ageing of the foam pushing the screen and the film together (Fig. 62). Dirt or grains may also cause artefacts. If cleaning the screen does not resolve the issue, it is possible to transfer the screen to a new cassette and reject the old cassette.

7.1.8. Light-tightness and safelight illumination of the darkroom

7.1.8.1. Description and objective

The light-tightness of the darkroom is important because a non-insulated darkroom can ruin all the films in a department or significantly degrade their

FIG. 62. Cassette with dried out foam causing artefacts because of bad film–screen contact.

quality. This test has two parts: one is used to confirm that safelight illumination is appropriate and the other confirms that light sources (safelight(s) and a non-light-tight darkroom) do not produce unacceptable fogging on the films.

This test may be repeated with each or every safelight on. In this case, the filter (quality and spectra) and its integrity can be evaluated for its effect on fogging and to confirm that the power rating of the light bulb is correct [4, 33].

7.1.8.2. Equipment

The following equipment is needed:

(a) The most sensitive film available (the film with the highest speed);
(b) A cardboard box;
(c) Six coins;
(d) A watch.

7.1.8.3. Procedure

The test procedure is as follows:

(a) Visual inspection:
    (i)   Step into the darkroom. Shade the windows, if any, and make the darkroom light-tight as usual.
    (ii)   Turn off all the lights in the darkroom, including the safelight.
    (iii) Allow sufficient time (approximately 5 min) for your eyes to adjust to the low light conditions.
    (iv)   Check for any spots letting white light into the room through the doors, processor, film exchange boxes, extractors or ceiling.
(b) Qualitative measurement:
    (i)   Load a film into a cassette. Always use a new package of films because a package opened earlier may be already foggy.
    (ii)   Expose half of the film to X rays, choosing exposure settings to obtain an optical density of 1 (sensitization exposure). The other half of the film may either be covered by a lead sheet or the film could be cut into half to observe the differences.
    (iii) Place the sensitized film on a workbench in the darkroom.
    (iv)   Place the six coins at approximately the same distance from each other along the longer side of the film.
    (v)   Use cardboard to cover the area of the film containing five of the coins. Wait for 30 s and move the cardboard to reveal one more coin. Repeat until the film and all coins are completely uncovered.
    (vi)   When the last 30 s have elapsed and none of the coins is covered, process the film.
    (vii) Repeat steps (i)–(vi) to test with any safelight illumination used.

7.1.8.4. Analysis and interpretation

Visually inspect the processed films for any images of coins.

7.1.8.5. Baselines and tolerances

No coins should be visible in the images.

7.1.8.6. Frequency

Repeat the test every six months and when a fault in the light-tightness of the darkroom is suspected.

The darkroom should be insulated against stray light and the safelight should be checked to ensure that it is not leaking white light. If needed, replace the filter of the safelight and check the power rating of the bulbs.

7.1.9.   X ray film viewer box

7.1.9.1.   *Description and objective*

The film viewer box should produce uniform light to illuminate the films. Inhomogeneous white light or flickering from the film viewer box might impact clinical evaluation of the films [4, 33].

7.1.9.2.   *Equipment*

An optical photometer calibrated for luminance measurements (in cd/m^2) is required for this test.

7.1.9.3.   *Procedure*

The test procedure is as follows:

(a)   Clean the viewer box and note any faulty or flickering lights inside.
(b)   Hold the photometer and measure the luminance of the film viewer at approximately 15 cm from the box on a 15 cm × 15 cm square area.
(c)   Record the measured values and then repeat step (b) as many times as necessary to cover the whole area of the X ray film viewer box.

7.1.9.4.   *Analysis and interpretation*

The homogeneity of the brightness should be evaluated by calculating the mean brightness of the sampled areas. This method may also be used to make a comparison of viewer boxes at a department.

7.1.9.5.   *Baselines and tolerances*

The brightness should be ±20% around the average of the measured values of the viewer box. No flickering or other artefacts should be observable (see Fig. 63).

FIG. 63. Flickering film viewer box (made with a very low shutter speed).

The baseline value for the average should be measured at commissioning (or soon after it). The typical value of luminance is usually in the range of several thousand candelas per square metre. However, to evaluate whether the viewing conditions are acceptable, the ambient lighting should also be evaluated.

7.1.9.6. Frequency

The test is repeated every three months.

7.1.9.7. Corrective actions

If the lights flicker or some of them are dim, a technician should replace them and check the power supply and the electronics. If the average value is significantly lower than the measured value at commissioning, then the glass (or plastic) plate in front of the lights should be thoroughly cleaned or replaced.

APPENDIX

This Appendix provides a list of the tests described in this handbook. While its aim is not to repeat content, the most important information related to each test is listed in Table 8: the section and title, the recommended equipment to be used, the suggested tolerance and frequency of each test, along with the bibliographical references serving as a basis for the given section.

TABLE 8. SUMMARY OF QUALITY CONTROL TESTS

Section	Title	Equipment	Tolerance	Frequency	Reference
Radiography — QC tests for radiographers					
2.1.1	X ray–light beam alignment and centring	Alignment test object, attenuator	Alignment: ±2% of SID Centring: ±1% of SID	3–6 months	[4, 5]
2.1.2	Distances and scales	Measuring tape	±1.5 cm	6 months	[4]
2.1.3	Image uniformity and artefacts	Cu plate or PMMA	No artefacts	Monthly	[4]
2.1.4	AEC constancy	Cu plate or PMMA	Digital radiography: ±25% of baseline Computed radiography: ±30% of baseline	3 months	[4, 6]
2.1.5	Condition of cassettes and image plates (computed radiography only)	Cleaning solution	No dirt or damage	According to the supplier's recommendations	
2.1.6	AEC sensitivity	Cu plate or PMMA	Exposure index or mAs value: ±50% of baseline	1–3 months	[4, 6]

TABLE 8. SUMMARY OF QUALITY CONTROL TESTS (cont.)

Section	Title	Equipment	Tolerance	Frequency	Reference
Radiography — QC tests for medical physicists					
2.2.1	X ray–light beam alignment and centring	Alignment test object, attenuator	Alignment: ±2% of SID Centring: ±1% of SID	3–6 months	[4, 5]
2.2.2	Tube potential accuracy	Solid state kilovolt meter, Pb plate	±5% or ±5 kV or ±10% or ±10 kV (IEC), whichever is greater	Annual	[4]
2.2.3	Radiation output consistency	Solid state detector or ionization chamber, Pb plate	±20% of baseline or 25–80 μGy/mA · s at 80 kV	Annual	[4, 7]
2.2.4	Short term reproducibility of radiation output and exposure time	Solid state detector or ionization chamber, Pb plate	COV of Y: ±20% of mean COV of t: ±5%	Annual	[4, 5, 7]
2.2.5	Exposure time accuracy	Solid state detector or ionization chamber, Pb plate	If $t > 100$ ms: ±10% of nominal If $t \leq 100$ ms: ±15% or ±2 ms of nominal	Annual	[4]

TABLE 8. SUMMARY OF QUALITY CONTROL TESTS (cont.)

Section	Title	Equipment	Tolerance	Frequency	Reference
2.2.6	Half-value layer	Solid state detector or ionization chamber, Al attenuators, metal plate, measuring tape	Compliance with minimum values in national regulations At 80 kV: >2.3 mm Al (X ray systems marketed before 2012-06-01) and 2.9 mm Al (X ray systems marketed after 2012-06-01)	Annual	[6, 8, 9]
2.2.7	KAP meter accuracy	Solid state detector or ionization chamber, plate or film, ruler or measuring tape, Pb plate	Standalone KAP meters: ±25% Otherwise: ±35%	Annual	[9–11]
2.2.8	Short term reproducibility of exposure indicator	Attenuator (e.g. Cu plate, PMMA)	±10% of baseline	Annual	[4, 12]
2.2.9	Exposure indicator accuracy	Cu plate or PMMA	±20% of baseline	Annual	[4, 6, 12]
2.2.10	Image receptor dose	Solid state detector or ionization chamber, attenuator (e.g. Cu plate, PMMA)	±30% of baseline	Annual	[4, 6]

TABLE 8. SUMMARY OF QUALITY CONTROL TESTS (cont.)

Section	Title	Equipment	Tolerance	Frequency	Reference
2.2.11	Leakage radiation	Survey meter, Pb sheet, ruler or measuring tape	1 mGy/h	At acceptance and when necessary	[14]
2.2.12	Scattered radiation	Survey meter, water phantom or PMMA, ruler or measuring tape	±50% of baseline	Annual	[15]
2.2.13	Low contrast detectability	Low contrast test object, attenuator (e.g. Cu plate, PMMA)	Decided by radiologist and radiographer	6 months	[13, 16]
2.2.14	Limiting spatial resolution	Bar pattern	No change from baseline	6 months	[4]
2.2.15	Dark noise	Pb sheets or Pb apron	±50% of baseline	Annual	[4, 13, 17]
2.2.16	Accuracy of measured dimensions	Test object or Pb ruler	±2%	Annual	[4, 13]
2.2.17	AEC consistency between sensors	Attenuator (e.g. PMMA slabs)	±30% of baseline ±20% of mean	Annual	[4, 13]
2.2.18	AEC system short term reproducibility	Attenuator (e.g. PMMA slabs)	±40% of mean	Annual	[4, 12]

TABLE 8. SUMMARY OF QUALITY CONTROL TESTS (cont.)

Section	Title	Equipment	Tolerance	Frequency	Reference
2.2.19	AEC system kilovoltage and thickness compensation	Attenuators (e.g. PMMA slabs) of different thicknesses	±40% of mean	Annual	[4, 12]
2.2.20	Operation of the AEC guard timer	Solid state detector or ionization chamber, Pb sheet	No change from baseline values	Annual	[4, 12]
2.2.21	Image uniformity and computed radiography plate sensitivity matching	Cu plate (1 mm thick) or PMMA (20 cm)	MPV: ±20% of mean mAs value: ±5% of mean No artefacts	Annual (computed radiography: 6 months)	[4, 12, 16]
2.2.22	Erasure efficiency (computed radiography only)	Pb sheet (2 mm thick) or Cu plate (>3 mm thick), Pb apron	No ghost image	Annual	[16]
Fluoroscopy — QC test for radiographers					
3.1.1	Reproducibility of AERC	PMMA (10–20 cm thick) or Cu plate (1–2 mm thick)	Kilovoltage: ±5% of baseline Current: ±20% of baseline	3 months	[7]
Fluoroscopy — QC tests for medical physicists					

188

TABLE 8. SUMMARY OF QUALITY CONTROL TESTS (cont.)

Section	Title	Equipment	Tolerance	Frequency	Reference
3.2.1	Verification of beam collimation	Phantom, computed radiography plate or self-developing film	±2% of SID	Annual	[14]
3.2.2	Verification of beam geometry	Alignment test objects, attenuator, computed radiography plate or self-developing film	Alignment: ±2% SID Centring: ±1% SID	Annual	[4, 5]
3.2.3	Verification of different field sizes	Phantom	±2% of SID	Annual	[14]
3.2.4	Patient ESAK rate	Ionization chamber or solid state detector, PMMA slabs	Manufacturer's specifications or 88 mGy/min (normal mode) 176 mGy/min (high dose rate mode)	Annual	[5, 6, 8, 9, 11]
3.2.5	Image receptor ESAK rate	Ionization chamber or solid state detector, 25 mm Al slab, Cu plates	0.2–1 µGy/s (with Al, 70–80 kV) ±25% of baseline	Annual	[4, 7]

TABLE 8. SUMMARY OF QUALITY CONTROL TESTS (cont.)

Section	Title	Equipment	Tolerance	Frequency	Reference
3.2.6	KAP meter calibration	Solid state detector or ionization chamber, plate or film, Pb plate	Standalone KAP meters: ±25% Otherwise: ±35%		
3.2.7	Leakage radiation	Survey meter, Pb sheet	1 mGy/h	At acceptance and when necessary	[13]
3.2.8	Scattered radiation	Survey meter, water phantom or PMMA	±50% of baseline	Annual	[14]
3.2.9	Image quality of fluoroscopy	Low contrast inserts, mesh/grid, spatial resolution test pattern, Cu step wedge	No change from baseline No artefacts	Annual	[4, 15]
3.2.10	Image quality of digital subtraction angiography	Phantom	No change from baseline No artefacts No logarithmic error (only for intensifiers)	Annual	[17]

TABLE 8. SUMMARY OF QUALITY CONTROL TESTS (cont.)

Section	Title	Equipment	Tolerance	Frequency	Reference
		Mammography — QC tests for radiographers			
4.1.1	Image receptor uniformity — weekly test (computed radiography and digital radiography)	PMMA slabs (45 mm thick) or manufacturer's phantom	±15% of MPV in whole image	Weekly	[18]
4.1.2	Subjective image quality evaluation	Breast mimicking phantom, baseline phantom image	mAs value: ±10% No significant image degradation No altered noise No artefacts No bright/dark pixels	Weekly	[18]
		Mammography — QC tests for medical physicists			
4.2.1	Compression force and thickness indicator accuracy	Scales, towels or blocks of rubber foam, PMMA slabs	Maximum automatic compression: 150–200 N Maximum manual force: 300 N Display accuracy: ±20 N Thickness: ±8 mm of slab thickness	Annual	[18]

TABLE 8. SUMMARY OF QUALITY CONTROL TESTS (cont.)

Section	Title	Equipment	Tolerance	Frequency	Reference
4.2.2	Detector alignment, alignment of X ray field to detector area	Computed radiography plate or self-developing film, radiopaque tools, phosphorescent strips, metal foil, PMMA slabs	Missing tissue on chest wall side: ±5 mm Otherwise: see Section 4.2.2.5	Annual and after service/ replacement	[18, 19]
4.2.3	Tube output	Solid state detector or ionization chamber, metal plate	None	Annual	[4, 18, 20]
4.2.4	Half-value layer	Ionization chamber, measuring tape, aluminium filters, metal plate	See Eq. (5)	Annual	[9, 18]
4.2.5	AEC reproducibility	PMMA slabs (45 mm thick), Al plate (0.2 cm thick), ionization chamber (for film–screen systems)	Computed radiography and digital radiography: ±5% COV of SDNR Film–screen systems: ±5% of incident air kerma	6 months	[18, 20, 21]

TABLE 8. SUMMARY OF QUALITY CONTROL TESTS (cont.)

Section	Title	Equipment	Tolerance	Frequency	Reference
4.2.6	AEC breast thickness compensation	PMMA slabs, Al plate, densitometer (for film–screen systems)	See Table 3	Annual	[18, 20–22]
4.2.7	AEC consistency between sensors	PMMA slabs, Al plate, ionization chamber (for film–screen systems)	Computed radiography and digital radiography: SDNR ±5% of mean Film–screen systems: OD = ±0.20 of mean	6 months	[18, 21]
4.2.8	Operation of AEC guard timer	Metal plate, ionization chamber or solid state detector (optional)	No change	At acceptance	[18, 21]
4.2.9	Response function and noise evaluation (computed radiography and digital radiography)	PMMA slabs (45 mm thick)	See Section 4.2.9.5	Annual	[18, 20]
4.2.10	Image receptor uniformity — annual test (computed radiography digital radiography)	PMMA slabs (45 mm thick)	Local uniformity: ±5% Global uniformity: ±10% (digital radiography only)	Annual	[20, 21]

TABLE 8. SUMMARY OF QUALITY CONTROL TESTS (cont.)

Section	Title	Equipment	Tolerance	Frequency	Reference
4.2.11	Spatial resolution	Straight edge metal attenuator, line pair resolution tool (for film–screen systems), PMMA slabs (45 mm thick), software	See Section 4.2.11.5	Annual	[18, 21]
4.2.12	Ghosting (computed radiography and digital radiography)	PMMA slabs (45 mm thick)	Ghost image SDNR ≤2.0	Annual and after detector replacement	[18]
4.2.13	Computed radiography plate sensitivity matching (computed radiography only)	PMMA slabs (45 mm thick)	MPV, SNR, exposure index: ±15% of mean and ±20% of the corresponding values of the reference image	Annual and for new cassettes	[18, 20]
4.2.14	Mean glandular dose	PMMA slabs, spacers, ionization chamber or solid state detector	See Table 4	Annual	[18, 20, 21]
4.2.15	Subjective evaluation of image quality	Phantom with low contrast objects, baseline phantom image	None	Annual	[18, 20, 22]

TABLE 8. SUMMARY OF QUALITY CONTROL TESTS (cont.)

Section	Title	Equipment	Tolerance	Frequency	Reference
		Computed tomography — QC tests for radiographers			
5.1.1	Daily startup procedure			Daily	[23]
5.1.2	CT alignment laser beams	Test device per Section 5.2.7	±5 mm	Monthly	[23]
5.1.3	Scan projection radiograph accuracy	Test object	±2 mm	6 months	[23]
5.1.4	CT number accuracy, image noise, image uniformity and image artefacts	Water phantom	CT number: ±5 HU of baseline (water) Noise: ±25% of baseline Uniformity: ±10 HU (water) No artefacts	6 months (CT number accuracy), weekly (other properties)	[23, 24]
5.1.5	Accuracy of measured dimensions	Test object	±2% of nominal	3 months	[4, 12]

TABLE 8. SUMMARY OF QUALITY CONTROL TESTS (cont.)

Section	Title	Equipment	Tolerance	Frequency	Reference
		Computed tomography — QC tests for medical physicists			
5.2.1	CT number accuracy, image noise, image uniformity and image artefacts	Test object	CT number: ±5 HU of baseline (water) Noise: ±25% of baseline Uniformity: ±10 HU (water) No artefacts	6 months	[23, 24]
5.2.2	Linearity	Test object	Water: ±4 HU of baseline Other materials: ±10 HU of baseline Radiotherapy, other materials: ±20 HU of reference	Annual	
5.2.3	Low contrast detail detectability	Test object	See Section 5.2.3.5	Annual and after changes	[23, 24]
5.2.4	X ray beam width	Test object	Greater of ±3 mm or ±30% of the total nominal collimated beam width	Annual and after changes	[23, 24]
5.2.5	Reconstructed image slice width	Test object	Table 5	Annual and after changes	[23]

TABLE 8. SUMMARY OF QUALITY CONTROL TESTS (cont.)

Section	Title	Equipment	Tolerance	Frequency	Reference
5.2.6	Spatial resolution	MTF test object, software, or bar pattern (alternatively)	Manufacturer's specifications	At acceptance and after changes	[23]
5.2.7	CT dosimetry	Pencil ionization chamber, CTDI phantom	C_{vol}: ±20% of baseline C_{air}: ±20% of baseline	Annual and after changes	[23–27]
Common aspects of modalities used in diagnostic radiology — QC tests for radiographers					
6.1.1	Routine check of image display	AAPM TG18-QC test pattern	Adjacent steps are distinguishable All low contrast inset patches are visible	Monthly	[23]
6.1.2	Reject rate assessment	Charts and logs	<8% <5% for paediatrics	1–3 months	[29]
6.1.3	Visual inspection of the workplace	Thermometer, checklist	See Section 6.1.3.5	Monthly and after changes	[18, 23]

TABLE 8. SUMMARY OF QUALITY CONTROL TESTS (cont.)

Section	Title	Equipment	Tolerance	Frequency	Reference
Common aspects of modalities used in diagnostic radiology — QC test for medical physicists					
6.2.1	Image display systems	AAPM TG18-QC test pattern	See Table 7	Annual	[23, 24, 31]
Film–screen systems — QC tests for radiographers					
7.1.1	Temperature of the film processing system	Digital thermometer (calibrated)	±1.0°C	Daily	[4, 22, 33]
7.1.2	Base and fog, sensitivity and contrast	Sensitometer, densitometer	OD = ±0.15 for general purpose films OD = ±0.10 for mammography OD = +0.03 for base and fog	Daily	[22]
7.1.3	pH measurement of the fixer solution	pH indicator strip and chart	Manufacturer's specification	Monthly	[4, 33]
7.1.4	Maintenance and visual inspection of cassettes	Marker pens, labels, cleaning solution, padding	Light-tight No defects	6 months and as needed	[4, 33]
7.1.5	Maintenance of screens	Cleaning solution, cloth, brush	No artefacts	Weekly and as needed	[4, 22, 33]

TABLE 8. SUMMARY OF QUALITY CONTROL TESTS (cont.)

Section	Title	Equipment	Tolerance	Frequency	Reference
7.1.6	Light-tightness of cassettes	Film viewer box, densitometer	No fogging	6 months and as needed	[4, 33]
7.1.7	Film–screen contact	Metallic mesh test object	No artefacts	6 months	[4, 33]
7.1.8	Light-tightness and safelight illumination of the darkroom	Films, cardboard box, coins, watch	No visible shadows	6 months and as needed	[4, 33]
7.1.9	X ray film viewer box	Photometer	±20% of mean No flickering	3 months	[4, 33]

Note: QC: quality control; SID: source–image distance; PMMA: polymethyl methacrylate; AEC: automatic exposure control; IEC: International Electrotechnical Commission; COV: coefficient of variation; KAP: kerma–area product; MPV: mean pixel value; AERC: automatic exposure rate control; SDNR: signal difference to noise ratio; CT: computed tomography; CTDI: CT dose index; Y: radiation output; t: exposure time; C_{vol}: volumetric CTDI; C_{air}: CTDI measured in air.

REFERENCES

[1] EUROPEAN COMMISSION, FOOD AND AGRICULTURE ORGANIZATION OF THE UNITED NATIONS, INTERNATIONAL ATOMIC ENERGY AGENCY, INTERNATIONAL LABOUR ORGANIZATION, OECD NUCLEAR ENERGY AGENCY, PAN AMERICAN HEALTH ORGANIZATION, UNITED NATIONS ENVIRONMENT PROGRAMME, WORLD HEALTH ORGANIZATION, Radiation Protection and Safety of Radiation Sources: International Basic Safety Standards, IAEA Safety Series No. GSR Part 3, IAEA, Vienna (2014).

[2] INTERNATIONAL ATOMIC ENERGY AGENCY, Radiation Comprehensive Clinical Audits of Diagnostic Radiology Practices: A Tool for Quality Improvement, IAEA Human Health Series No. 4, IAEA, Vienna (2010).

[3] WORLD HEALTH ORGANIZATION, Quality Assurance in Diagnostic Radiology, WHO, Geneva (1982).

[4] INSTITUTE OF PHYSICS AND ENGINEERING IN MEDICINE, Recommended Standards for the Routine Performance Testing of Diagnostic X-ray Imaging Systems, IPEM Report 91, IPEM, York (2005).

[5] AMERICAN ASSOCIATION OF PHYSICISTS IN MEDICINE, Quality Control in Diagnostic Radiology, AAPM Report No. 74, AAPM, New York (2002).

[6] EUROPEAN COMMISSION, Criteria for Acceptability of Medical Radiological Equipment Used in Diagnostic Radiology, Nuclear Medicine and Radiotherapy, Radiation Protection No. 162, EC, Luxembourg (2013).

[7] INTERNATIONAL ELECTROTECHNICAL COMMISSION, Evaluation and Routine Testing in Medical Imaging Departments — Part 3-1: Acceptance Tests — Imaging Performance of X-ray Equipment for Radiographic and Radioscopic Systems, IEC 61223-3-1, IEC, Geneva (1999).

[8] INTERNATIONAL ELECTROTECHNICAL COMMISSION, Medical Electrical Equipment — Part 1-3: General Requirements for Basic Safety and Essential Performance — Collateral Standard: Radiation Protection in Diagnostic X-ray Equipment, IEC 60601-1-3, IEC, Geneva (2008).

[9] INTERNATIONAL ATOMIC ENERGY AGENCY, Dosimetry in Diagnostic Radiology: An International Code of Practice, Technical Report Series No. 457, IAEA, Vienna (2007).

[10] INTERNATIONAL ELECTROTECHNICAL COMMISSION, Medical Electrical Equipment — Dose Area Product Meters, IEC 60580, IEC, Geneva (2019).

[11] INTERNATIONAL ELECTROTECHNICAL COMMISSION, Medical Electrical Equipment — Part 2–54: Particular Requirements for the Basic Safety and Essential Performance of X-ray Equipment for Radiography and Radioscopy, IEC 60601-2-54, IEC, Geneva (2009).

[12] INSTITUTE OF PHYSICS AND ENGINEERING IN MEDICINE, Measurement of the Performance Characteristics of Diagnostic X-ray Systems: Digital Imaging Systems, IPEM Report 32 (VII), IPEM, York (2010).

[13] EUROPEAN COMMISSION, Criteria for Acceptability of Radiological (Including Radiotherapy) and Nuclear Medicine Installations, Radiation Protection No. 91, EC, Luxembourg (1997).

[14] INTERNATIONAL ELECTROTECHNICAL COMMISSION, Medical Electrical Equipment — Part 2-43: Particular Requirements for the safety of X-ray equipment for Interventional Procedures, IEC 60601-2-43, IEC, Geneva (2011).

[15] DEUTSCHES INSTITUT FÜR NORMUNG, Image Quality Assurance in Diagnostic X-ray Departments — Part 4: Constancy Testing of Medical X-ray Equipment for Fluoroscopy, DIN 6868-4, DIN, Berlin (2007).

[16] AMERICAN ASSOCIATION OF PHYSICISTS IN MEDICINE, Acceptance Testing and Quality Control of Photostimulable Phosphor Imaging Systems, Report No. 93, AAPM, College Park, MD (2006).

[17] INTERNATIONAL ELECTROTECHNICAL COMMISSION, Evaluation and Routine Testing in Medical Imaging Departments — Part 3-3: Acceptance tests — Imaging Performance of X-ray Equipment for Digital Subtraction Angiography (DSA), IEC 61223-3-3, IEC, Geneva (2010).

[18] INTERNATIONAL ATOMIC ENERGY AGENCY, Quality Assurance Programme for Digital Mammography, IAEA Human Health Series No. 17, IAEA, Vienna (2011).

[19] INSTITUTE OF PHYSICAL SCIENCES IN MEDICINE, The Commissioning and Routine Testing of Mammographic X-ray systems: A Protocol Produced by a Working Party of the Diagnostic Radiology Topic Group, Report No. 59, IPSM, York (1995).

[20] EUROPEAN FEDERATION OF ORGANISATIONS FOR MEDICAL PHYSICS, Mammo Protocol, EFOMP, Utrecht (2015),
https://www.efomp.org/index.php?r=fc&id=protocols

[21] EUROPEAN COMMISSION, DIRECTORATE-GENERAL FOR HEALTH AND CONSUMERS, European Guidelines for Quality Assurance in Breast Cancer Screening and Diagnosis, 4th edn (VON KARSA, L., et al., Eds), EC, Luxembourg (2006).

[22] INTERNATIONAL ATOMIC ENERGY AGENCY, Quality Assurance Programme for Screen Film Mammography, IAEA Human Health Series No. 2, IAEA, Vienna (2009).

[23] INTERNATIONAL ATOMIC ENERGY AGENCY, Quality Assurance Programme for Computed Tomography: Diagnostic and Therapy Applications, IAEA Human Health Series No. 19, IAEA, Vienna (2012).

[24] AMERICAN COLLEGE OF RADIOLOGY, Computed Tomography Quality Control Manual, ACR (2017),
https://www.acr.org/-/media/ACR/Files/Clinical-Resources/QC-Manuals/CT_QCManual.pdf

[25] INTERNATIONAL ELECTROTECHNICAL COMMISSION, Medical Electrical Equipment — Part 2-44: Particular Requirements for the Basic Safety and Essential Performance of X-ray Equipment for Computed Tomography, IEC 60601-2-44, IEC, Geneva (2009).

[26] INTERNATIONAL ELECTROTECHNICAL COMMISSION, Medical Electrical Equipment — Dosimeters with Ionization Chambers and/or Semiconductor Detectors as Used in X-ray Diagnostic Imaging, IEC 61674, IEC, Geneva (2012).

[27] INTERNATIONAL ATOMIC ENERGY AGENCY, Status of Computed Tomography Dosimetry for Wide Cone Beam Scanners, IAEA Human Health Reports No. 5, IAEA, Vienna (2011).

[28] SAMEI, E., et al., Assessment of display performance for medical imaging systems: Executive summary of AAPM TG18 report, Med. Phys. **32** (2005) 1205–1225.

[29] JONES, A.K., Ongoing quality control in digital radiography: Report of AAPM Imaging Physics Committee Task Group 151, Med. Phys. **42** (11), American Association of Physicists in Medicine, Alexandria (2015).

[30] BEVINS, N.B., et al., Display Quality assurance – The Report of AAPM Task Group 270, AAPM Report No. 270, American Association of Physicists in Medicine, Alexandria (2019),
https://www.aapm.org/pubs/reports/RPT_270.pdf

[31] AMERICAN COLLEGE OF RADIOLOGY, AMERICAN ASSOCIATION OF PHYSICISTS IN MEDICINE, SOCIETY FOR IMAGING INFORMATICS IN MEDICINE, ACR–AAPM–SIIM Technical Standard for Electronic Practice of Medical Imaging, ACR (2017),
https://www.acr.org/-/media/ACR/Files/Practice-Parameters/elec-practice-medimag.pdf

[32] GRESS, D.A., et al., AAPM medical physics practice guideline 6.a.: Performance characteristics of radiation dose index monitoring systems, J. Appl. Clin. Med. Phys. **18** (2017) 12–22.

[33] FINCH, A., BURY, R., WORKMAN, A., Assurance of Quality in the Diagnostic Imaging Department: Prepared by the Quality Assurance Working Group of the Radiation Protection Committee of the British Institute of Radiology, 2nd edn, British Institute of Radiology, London (2001).

ABBREVIATIONS

AEC	Automatic exposure control
AERC	Automatic exposure rate control (identical to automatic brightness control)
COV	Coefficient of variation
CT	Computed tomography
CTDI	Computed tomography dose index
DMS	Dose management software
ESAK	Entrance surface air kerma
FOV	Field of view
HVL	Half-value layer
IEC	International Electrotechnical Commission
ISRRT	International Society of Radiographers and Radiological Technologists
KAP	Kerma–area product
MGD	Mean glandular dose
MPV	Mean pixel value
MTF	Modulation transfer function
PMMA	Polymethyl methacrylate
QA	Quality assurance
QC	Quality control
ROI	Region of interest
SDNR	Signal difference to noise ratio
SID	Source to image distance
SNR	Signal to noise ratio
SPR	Scan projection radiograph

CONTRIBUTORS TO DRAFTING AND REVIEW

Arreola, M.	University of Florida, United States of America
Brambilla, M.	Azienda Ospedaliero-Universitaria Maggiore della Carità, Italy
Burinskiene, V.	Kaunas Medical University Hospital, Lithuania
Ciraj-Bjelac, O.	International Atomic Energy Agency
Coskun, N.	Manisa State Hospital, Türkiye
Delis, H.	International Atomic Energy Agency
Elek, R.	National Public Health Centre, Hungary
Faj, D.	J.J. Strossmayer University, Croatia
Faulkner, K.	Public Health England, United Kingdom
Gazdic-Santic, M.	Clinical Centre in Sarajevo, Bosnia and Herzegovina
Gershan, V.	University Ss Cyril and Methodius, North Macedonia
Giannos, A.	Bank of Cyprus Oncology Centre, Cyprus
Gingold, E.L.	Thomas Jefferson University, United States of America
Ivanovic, S.	Clinical Centre of Montenegro, Montenegro
Kalathaki, M.	Greek Atomic Energy Commission, Greece
Kaplanis, P.A.	Ministry of Health, Cyprus
Katsifarakis, D.	International Society of Radiographers and Radiological Technologists, Greece
Kishta, D.	University Hospital Center "Mother Theresa", Albania
Kjakste, I.	Riga 2nd Hospital, Latvia
Kostova-Lefterova, D.	National Centre of Radiobiology and Radiation Protection, Bulgaria

Krylova, T.A.	Russian Cancer Research Centre, Russian Federation
Mirkov, Z.	Serbian Institute of Occupational Health, Serbia
Porubszky, T.	National Public Health Centre, Hungary
Ribeiro, T.N.R.	Centro Hospitalar Lisboa Norte, Portugal
Ruuge, P.	East Tallinn Central Hospital, Estonia
Tsapaki, V.	International Atomic Energy Agency
Tsitovich, Y.	Research Institute of Oncology and Medical Radiology, Belarus
Vassileva, J.	International Atomic Energy Agency
Whitley, A.S.	International Society of Radiographers and Radiological Technologists, Greece

Consultants Meetings

Vienna, Austria: 18–22 April 2016, 12–16 June 2017

 **IAEA**
International Atomic Energy Agency

ORDERING LOCALLY

IAEA priced publications may be purchased from the sources listed below or from major local booksellers.

Orders for unpriced publications should be made directly to the IAEA. The contact details are given at the end of this list.

NORTH AMERICA

Bernan / Rowman & Littlefield
15250 NBN Way, Blue Ridge Summit, PA 17214, USA
Telephone: +1 800 462 6420 • Fax: +1 800 338 4550
Email: orders@rowman.com • Web site: www.rowman.com/bernan

REST OF WORLD

Please contact your preferred local supplier, or our lead distributor:

Eurospan Group
Gray's Inn House
127 Clerkenwell Road
London EC1R 5DB
United Kingdom

Trade orders and enquiries:
Telephone: +44 (0)176 760 4972 • Fax: +44 (0)176 760 1640
Email: eurospan@turpin-distribution.com

Individual orders:
www.eurospanbookstore.com/iaea

For further information:
Telephone: +44 (0)207 240 0856 • Fax: +44 (0)207 379 0609
Email: info@eurospangroup.com • Web site: www.eurospangroup.com

Orders for both priced and unpriced publications may be addressed directly to:
Marketing and Sales Unit
International Atomic Energy Agency
Vienna International Centre, PO Box 100, 1400 Vienna, Austria
Telephone: +43 1 2600 22529 or 22530 • Fax: +43 1 26007 22529
Email: sales.publications@iaea.org • Web site: www.iaea.org/publications